THE FEAR FACTOR

THE FEAR FACTOR

How One Emotion
Connects Altruists, Psychopaths,
and Everyone In-Between

ABIGAIL MARSH

BASIC BOOKS
New York

Basic Books
Hachette Book Group
1290 Avenue of the Americas, New York, NY 10104
www.basicbooks.com

First edition: October 2017

Published by Basic Books, an imprint of Perseus Books, LLC, a subsidiary of Hachette Book Group, Inc.

Print book interior design by Jeff Williams

The Library of Congress has cataloged the hardcover edition as follows:

Names: Marsh, Abigail, author.
Title: The fear factor : how one emotion connects altruists, psychopaths, and everyone in-between / Abigail Marsh.
Description: First Edition. | New York : Basic Books, 2017. | Includes bibliographical references and index.
Identifiers: LCCN 2017014787 (print) | LCCN 2017022393 (ebook) ISBN 9781541697195 (hardback) | ISBN 9781541697201 (ebook) |
Subjects: LCSH: Fear. | Philanthropists—Psychology—Case studies. | Psychopaths—Psychology—Case studies. | BISAC: PSYCHOLOGY / Psychopathology / General. | SCIENCE / Life Sciences / Neuroscience.
Classification: LCC BF575.F2 (ebook) | LCC BF575.F2 M374 2017 (print) | DDC 152.4/6—dc23
LC record available at https://lccn.loc.gov/2017014787

ISBNs: 978-1-54169-719-5 (hardcover); 978-1-54169-720-1 (e-book)

LSC-C

10 9 8 7 6 5 4 3 2 1

To the man whose astonishing bravery
and compassion inspired this book, and to all the
other altruists whose actions have inspired those
whose lives they have touched.

CONTENTS

PROLOGUE

> It is extremely doubtful whether the offspring of the more sympathetic and benevolent parents, or of those which were the most faithful to their comrades, would be reared in greater number than the children of selfish or treacherous parents of the same tribe. He who was ready to sacrifice his life, as many a savage has been, rather than betray his comrades, would often leave no offspring to inherit his noble nature.
>
> —CHARLES DARWIN, *The Descent of Man*

> My favorite part is that other people say, "Oh, I could never do that." Well, that's bullshit.
>
> —*Altruistic kidney donor* HAROLD MINTZ, *on donating a kidney*

IN 1934, THE French entomologist Antoine Magnan set forth to write a scholarly text on the flight of insects. He ran into one niggling problem. After running calculations with an engineer named André Saint-Lagué, Magnan concluded that, according to the laws of aerodynamics, insects should not be able to fly at all. With a note of dejection, he wrote: "I applied the laws of air resistance to insects, and I arrived with Mr. St. Lagué at the conclusion that their flight is impossible."

And yet insects fly.

Conspiracy theorists love to use this apparent contradiction (sometimes touted as applying only to bees) to declare physics and biology to be bankrupt pursuits. Some religious devotees proclaim it as evidence of a higher power. But scientists are patient, and time is on their side.

Upon reading Magnan's assertion, entomologists did not decide that insect flight must be an illusion, or that it results from supernatural forces. Nor did they conclude that the laws of aerodynamics are hopeless bunk. They knew that reconciliation must be possible, but awaited better methods for measuring the properties of insect flight and calculating the physical dynamics at play.

Several decades and the invention of high-speed photography later, the puzzle has been solved. Insects, bees included, fly because their wings beat very quickly—bee wings make 230 short, choppy strokes every second—while rotating around their hinge to carve figure-eights in the air. The rotation creates a bug wing–sized vortex that generates enough lift to support a fat bug body. A robotic wing can be programmed to work precisely the same way, conclusively demonstrating that insect flight and the laws of physics are compatible.

Another apparent contradiction of the laws of nature, one that is arguably even more puzzling than insect flight, is altruism.

The theory of evolution by natural selection is as rock-solid as scientific laws get. But as Charles Darwin, the father of the theory, calculated some 150 years ago, natural selection seems to dictate that all the altruists should have died out long ago. An individual who sacrifices to help another person will do wonders for the other person's odds of survival, but not much for his own. Over the course of human history, the saps who sacrificed their own evolutionary fitness for others should have been outcompeted, outnumbered, and eventually completely superseded by their self-serving brethren.

And yet altruism exists.

I know this from personal experience. When I was nineteen, an altruistic stranger saved my life, gaining nothing in exchange for the risks he undertook to rescue me. And he was just one of many. Carnegie Hero Fund Medals are awarded every year to dozens of Americans who risk their lives to extraordinary degrees to save the lives of strangers. Over one hundred Americans a year undergo surgery, at no small risk to themselves, to donate a kidney to a stranger, often anonymously. Millions of people around the world donate bone marrow or blood—smaller sacrifices, certainly, but with no less noble a purpose: to help a stranger in need.

Until recently, there was no clear scientific explanation for actions like these. Since Darwin's era, biologists have developed models to explain altruistic behavior, but these models focus on altruism aimed at helping close kin or members of one's own social group. For example, some altruism toward kin can be explained via *inclusive fitness*. Inclusive fitness dictates that an altruistic behavior can evolve if its beneficiary shares enough genes with the altruist to compensate the altruist for the risk he or she has taken. It explains why colony-dwelling creatures like ground squirrels sound an alarm when a predator approaches. These calls attract the predator's attention and put the caller at risk, but they also help close relatives in the colony escape danger. Inclusive fitness may also explain why humans prefer to donate organs to close family members instead of strangers or friends. If you donate a kidney to your sister and she goes on to bear your nieces and nephews, they will carry some of your genes into the next generation. You may not personally benefit from your generosity, but your genes will, which makes the risk worthwhile from an evolutionary point of view.

What about altruism toward distantly related or unrelated others? Some such altruism takes the form of *reciprocal altruism*, which relies on the expectation that the beneficiary will one day return the favor. For example, vampire bats famously regurgitate blood into the mouths of even unrelated colony-mates who cannot find food and are at risk of starving. Their generosity pays off, though. Bats are more likely to receive blood buffets in the future from those with whom they have shared in the past. Humans engage in similar acts of reciprocity all the time, minus the regurgitation. You have probably lent a neighbor sugar or bought your colleagues coffee with the expectation that they will eventually reciprocate. Reciprocal altruism nearly always benefits members of the altruist's social group, who are more likely to be willing and able to return the favor later than would a passing stranger. This form of altruism is really a form of delayed gratification because ultimately the altruist will personally benefit, but only after some time has passed.

Both kin-based altruism and cooperation-based altruism are widespread and valuable biological strategies. Life as a social species would probably be impossible without them. Many books about altruism explore these forms of altruism in great detail. But both of these types of altruism are fundamentally selfish, in a sense. Kin-based altruism is directly aimed at benefiting the altruist's genes, and cooperation-based altruism is directly aimed at benefiting the altruist personally. So both of these models are useless for explaining the kind of altruism exhibited by altruistic kidney donors or Carnegie Heroes or the man who rescued me. These altruists intentionally and voluntarily risk their lives to save, not a relative or a friend, but an anonymous stranger. And they do it with no possible commensurate payoff to themselves, either genetically or personally. Indeed, they often pay dearly for their sacrifices. What could possibly explain their actions?

As in the case of insect flight, the seeming contradiction between altruism and the known laws of science often leads people to seek other explanations. Some declare all altruism an illusion. No matter how altruistic an action appears to be, no matter how great the risk and how small any possible payoff, perhaps it is really self-interest in disguise. Perhaps heroic rescuers are just looking for a rush and kidney donors are seeking public adulation. Others cite supernatural forces, calling heroic rescuers "guardian angels," or altruistic kidney donors "saints." Metaphorical or not, these terms suggest that whatever motivates these altruists cannot be explained by science. But scientists are patient, and time is on their side.

An avalanche of new technologies for studying human psychology and behavior have emerged in recent decades, including new methods for measuring and manipulating activity inside the brain, acquiring genetic information, and comparing human and animal behaviors. Much of this work has emerged at the intersection of established disciplines, spawning entirely new fields like social neuroscience and cognitive neurogenetics. Just as high-speed photography and robotics yielded new answers about insect flight, so has this profusion of technologies yielded new answers about human altruism.

My own rescue inspired me to take advantage of these new approaches to understand the origins of altruism. I was a college student at the time, and shortly thereafter I turned my academic focus to the study of psychology. I first conducted laboratory-based research as an undergraduate at Dartmouth College and later as a doctoral student at Harvard University. While working on my dissertation at Harvard, I made a serendipitous discovery. Efforts to find markers of highly altruistic people in the laboratory had until that point mostly failed. But I discovered that altruism is robustly related to how attuned people are to others' fear. People who can accurately label photos of frightened faces are also the people who donate the most money to a stranger under controlled laboratory conditions, or volunteer the most time to help them. The ability to label others' fear predicts altruism better than gender, mood, or how compassionate study participants claim to be, and this relationship holds up in study after study. But the question persisted: why?

Answers began to emerge as I continued my research in the laboratory of Dr. James Blair at the National Institute of Mental Health (NIMH). I joined the Bethesda, Maryland, lab just as it was embarking on the first-ever series of brain imaging studies to probe what makes psychopathic adolescents tick. This required using functional magnetic resonance imaging (fMRI) to scan the brains of teenagers at risk of becoming psychopaths. The results revealed that these teenagers' brains were marked by dysfunction in a structure called the *amygdala,* which is buried deep in the brain's interior and is responsible for essential social and emotional functions. In these teenagers who showed little empathy or compassion for others, the amygdala was underresponsive to images of others' fear. Moreover, this pattern of dysfunction seemed to prevent the teenagers from identifying fearful expressions. If amygdala dysfunction robs people of both

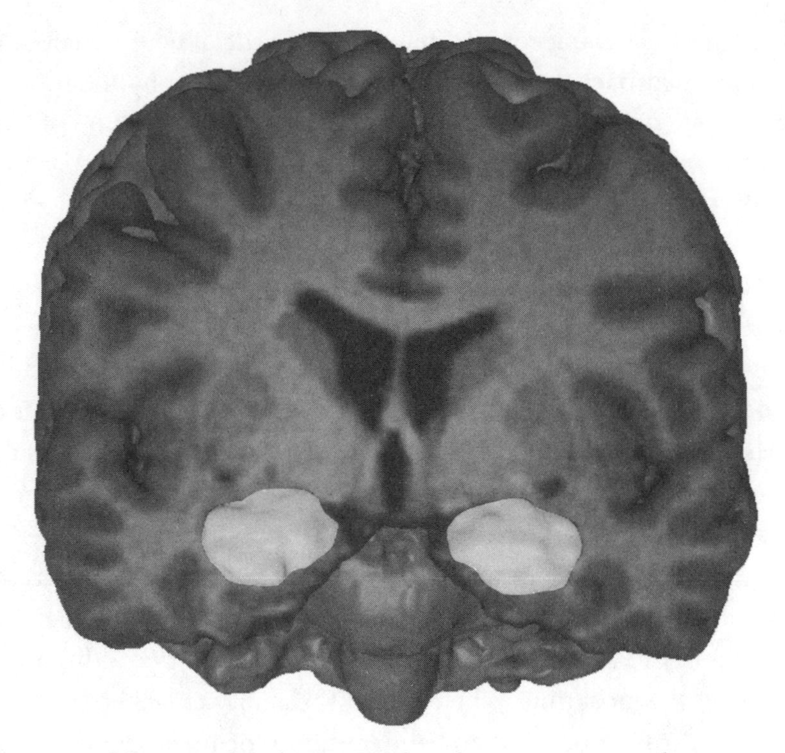

An image of the amygdala from one of our brain imaging studies.
Abigail Marsh and Katherine O'Connell.

empathy and the ability to recognize fear, could amygdala-based sensitivity to others' fear be a critical ingredient for altruism—including acts of extraordinary altruism like the one that saved my life?

Finding an answer would require locating real altruists and scanning their brains, which had never been previously attempted. Upon completing my postdoctoral research fellowship at NIMH, I began a professorship at Georgetown University, whereupon my research group set about recruiting nineteen altruistic kidney donors who had donated a kidney to a stranger. Some of them had responded to flyers posted by strangers seeking kidneys, and others had called a local transplant center and offered to give a kidney anonymously to anyone who needed it, no questions asked. None received payment in exchange for the inconvenience, pain, and small risk of serious injury or death that the surgery entailed. They weren't even compensated for the days of work they missed or their travel expenses. On the surface, these extraordinary altruists had little in common—they were men and women of varying ages, religious backgrounds, and political persuasions who came from all over America and told very different stories about what drove them to donate. But our research demonstrated that they did share something in common: an unusually strong amygdala response to pictures of other people's fear, as well as an enhanced ability to recognize it.

The Fear Factor delves deep inside the human brain to explore why sensitivity to others' fear is such a powerful marker for altruism, on the one hand, and for psychopathy, on the other. Findings from my own research, coupled with emerging knowledge from brain imaging and genetic studies, have provided new insights into the origins of empathy, psychopathy, and altruism. This book considers the question of how our species came to be endowed with the capacity to care by tracing altruism in modern humans back to the emergence of Earth's first mammals, who developed a desire to nurture and protect their offspring rather than let them fend for themselves. This desire springs in part from a chemical called *oxytocin*. Oxytocin is expressed strongly in the amygdala and may be able to convert the desire to avoid the distress of others into the desire to ameliorate it. New evidence suggests that psychopathy may result from breakdowns in the brain processes that evolved to enable parenting.

With this in mind, my NIMH colleagues and I developed a protocol to administer oxytocin intranasally to a sample of typical human research participants who came to the sprawling clinical center of the NIH. We evaluated how administering oxytocin affected the deep-seated social processes that underlie the capacity for altruism, like sensitivity to others' emotions and responses to the faces of infants. To put our findings in context, I tracked down stories of modern mammals from around the globe, from lions to golden retrievers, who have engaged in acts of extraordinary parenting. Understanding how fearsome carnivores like lions and dogs could be moved to nurture and protect creatures, like antelopes and squirrels, which they would normally hunt and kill, may hold the key to understanding equally unlikely acts of altruism in humans—and how to foster them. *The Fear Factor* considers whether, if the lion can lie down with the antelope (if not the lamb), we humans can learn to become more altruistic toward one another as well—and whether we should.

1

THE RESCUE

OVER BREAKFAST THE morning after the rescue my mom could tell just by looking at me that something had happened. I left it at, "I hit a dog on the freeway." Which was true. But the rest, the part I couldn't bring myself to tell her for fear she might lose her mind from retroactive panic, is the real start of my story.

I was driving home to Tacoma, Washington, after spending the evening with a childhood friend in Seattle. It was a clear summer night around midnight, traffic was light, and I was sober. All that was the fortunate part. Less fortunate was the car I was driving—my mom's SUV of a vintage that is infamous for being unstable during sharp turns. Normally Interstate 5 doesn't require any sharp turns as it wends its way from Seattle toward downtown Tacoma. The eight-lane freeway curves only slightly as it rises up over the Puyallup River, the peeling blue mass of the Tacoma Dome looming ahead of the southbound drivers.

I don't know where the dog came from. The overpass traverses an industrial area with no homes that a dog could have escaped from, and it has no shoulders that a dog could run along. It's hard to imagine a less likely place to run into a dog. But run into it I did. Or rather, I ran over it. I tried to avoid it, jerking the steering wheel as soon as I saw the tiny orangeish mass streaking across the road at a speed only an absolutely terrified dog could muster. Jerking the steering wheel to avoid an animal is, of course, the wrong response. Just mow it down is what all the experts say. But my first instinct was to avoid the dog, and I had no time to override that instinct. I love dogs. I wanted a dog so badly when I was in grade school that I daydreamed about being blind so that I could get a guide dog. Recalling the feel of my front wheel rising slightly as it rolled over that poor creature still makes me shudder.

What came next was worse. The combination of the car turning sharply and then tilting as it rolled over the dog destabilized the SUV and sent it into a fishtail. It veered sickeningly leftward across two lanes, then swooped back across them to the right as I fought for control of the wheel. But by the third turn the pull of the wheel had become too strong and I lost control completely. The car started to spin. A sickening succession of images passed before me as the car carved circles across the freeway: guardrail . . . headlights . . . guardrail . . . taillights . . . guardrail . . . and . . . headlights. Then, still facing the headlights of the oncoming cars, it slid to a stop.

Getting my bearings, I realized that I was in the far left lane of the freeway—the fast lane. Only now it was the far right lane because I was facing backward toward the oncoming traffic. Because the car had come to rest just past the apex of the overpass, it wasn't visible to drivers coming toward me until they crested the hill, which left them with little time to avoid me. Some of them came so close before they swerved that the car shook as they blew past.

There was no shoulder to escape onto. The bridge was hemmed in by guardrails and only inches separated them from where I sat. I couldn't have driven the car onto a shoulder anyhow because the engine had died. *Does spinning around kill a car's engine?* I remember wondering vaguely as I stared at a dashboard full of warning lights, including the ominous CHECK ENGINE. My teenage brother and his friends spent many a snowy day turning doughnuts in parking lots, but I didn't remember hearing that it made their engine die.

I turned the key in the ignition again and again, willing it to catch, but it stayed stubbornly silent. I knew that if I didn't do something it was only a matter of time until a car, or worse, one of the eighteen-wheelers barreling by, plowed into me. But what could I do? I turned on the brights. I turned on the emergency flashers. I had no mobile phone in 1996, so I couldn't call for help. Should I get out of the car and shimmy down the narrow shoulder away from it? And then what? Run across the freeway to an exit ramp? Or should I stay inside where at least I was protected by layers of metal and fiberglass and airbags?

I don't know how long I sat there, terrified, weighing these equally unappealing options. Thinking you're about to die makes time bunch up and stretch out in odd ways. But next thing I knew, I heard a rap on the half-open passenger's side window, on the side of the car nearest the guardrail.

I turned toward the sound and flinched when I saw who was standing there gazing in at me. I wasn't sure whether my situation had improved or just gotten much worse. Tacoma's downtown was racked by violence back then. The epicenter of the violence was the notorious Hilltop neighborhood just to the west of where I sat. The emergency rooms of the two hospitals that stood like concrete sentries on either side of the Hilltop treated a steady stream of shooting victims, most of them either members or rivals of the Hilltop Crips. People from the part of Tacoma where I lived did not hang out near the Hilltop. We definitely did not go there and interact with men like this one. He seemed an unlikely candidate for a roadside hero. He was wearing sunglasses, despite it being the middle of the night, and an abundance of gold jewelry. His head was shaved and shone like a coffee bean. When he spoke, I thought I saw the gleam of a gold tooth.

"You look like you could use some help," he said. His voice was low and rumbling.

"Um. I think I do," I responded, my voice catching in my throat.

"All right. Then I need to get in your seat." He gestured to the driver's seat.

Oh Jesus, I thought. *Now what? This man wants to get in the car with me?* My mom didn't even want my friends driving her car (understandably). What would she think of this man driving it? But I had zero other viable choices.

After a pause, I nodded. "Okay."

He walked around the hood of the car and watched the traffic for a moment. His head bobbed faintly leftward as each car passed, like a jump roper finding the rhythm. When a gap in the traffic appeared, he moved fast. In an instant, he was outside my door and yanking it open. I lifted myself over the center console to the passenger's side in time for him to swing himself into the driver's seat and slam the door shut behind him. He grasped the wheel and turned the key. Nothing.

"It won't start," I said. He turned the key again. Still nothing.

His gaze moved systematically across the dashboard and controls. It landed on the gearshift, which was still in drive. Without comment, he moved it back to park, then tried the ignition again. It caught! He pulled the gearshift back to drive, watched again for a gap in the flow of traffic, and, when one appeared, floored the accelerator to launch us in a smooth arc across the freeway. A moment later we were safe again—relatively speaking—on the diagonal stripes of the off-ramp. He eased to a stop behind his own car, a dark-colored BMW that gleamed orange under the sodium lights. Since getting into my car, he had not looked at me or spoken. Now he turned to me, taking in my jagged breathing and my face, which felt tight and drawn. My skin felt cold, and my legs were shaking uncontrollably.

"You going to be all right getting home? Need me to follow you for a bit?" he asked.

I shook my head. "No, I'll be all right. I can get home," I said.

Did I thank him? I'm not certain. I think I somehow forgot.

"Okay. You take care of yourself then," he said. And then he was gone. He returned to his car, and the summer night swallowed him up.

⁂

I never learned his name. I know nothing about him. Was he a Hilltop Crip, or was he an expensively dressed attorney or preacher or salesman? All I actually know about him is that while he was driving on Interstate 5 near midnight—Was he tired? Did he have somewhere to be?—he encountered an SUV marooned in the fast lane facing backwards, brights blazing and flashers blinking. Could he have even seen me inside? If he did, it could only have been for the briefest of moments. Most of the drivers passing by barely had time to veer out of the way. But in the second that passed between when he saw me and when he pulled over, he made an incredible decision: to try to save my life. He pulled into the off-ramp, parked, then ran some fifty feet across four lanes of the busiest freeway in Washington State, in the dark, to reach me. Did he have second thoughts as he stood staring at the cars and trucks blazing past him at freeway speed? If so, he didn't yield to them.

He then tested his luck against the traffic twice more: once to get around to the driver's side of the car, and then again to launch the car across the freeway from a dead stop. Any of those three times a miscalculation or a stroke of bad luck could have caused him—and maybe me—to die a violent death. But he did it anyhow. He did it to help me, a woman who found him frightening and couldn't pull herself together enough to thank him. He was clearly capable of great bravery and great selflessness. He couldn't have been looking for a reward—not even the recognition of a little story in the *Tacoma News Tribune*. By not telling me his name, he guaranteed that no one would ever know what he had done. He was a hero in every sense, and I am sorry to this day that I can't tell him that, and thank him for my life.

In the immediate aftermath of that night, I was mostly tormented by fear and regret: How did I manage not to hit another car as I was spinning across the freeway? What would have happened to me if the stranger had not arrived? Would I be lying mangled in an intensive care unit in a hospital on the Hilltop? Would I be dead?

My stomach turned every time I thought about the dog that had set off the whole chain of events. Such a frightened, helpless creature, and through my own stupid actions—intended, with terrible irony, to spare its life—I had killed it. Had it suffered? I hoped not. But I couldn't erase it from my mind. For weeks afterward I could still see matted, orange tufts of its fur stuck to the asphalt when I drove across the bridge over the Puyallup River.

As time went on, though, a new kind of torment took hold. It wasn't so much an emotional torment as an intellectual one. I found myself turning over and over in my head questions about my rescuer and the improbability of what he had done.

He wasn't a singularity, of course. Heroic rescues happen everywhere with some regularity. The Carnegie Hero Fund honors dozens of them every year. Everyone has heard news stories about someone who leapt into a river to save a drowning child or rushed into a burning building to rescue an old woman. But these stories are somehow remote and bloodless. It is easy, reading them, to discount the risks and pain that the rescuers faced—the freezing, rushing waters of the river, the heat and hiss of the flames—and the feelings they must have experienced. For most people, even those who can picture the scene vividly, the outlandishness of the rescuers' decision makes it difficult to imagine what could have been going through their heads. What were they feeling? Were they frightened? If so, how did they overcome their fear to act so bravely? The difficulty of comprehending the decision to risk suffering or death to save a stranger makes it tempting, I think, to treat the mind that would make such a decision as a locked box, remote and unknowable, and somehow fundamentally different from the minds of the rest of us.

I was stuck in the same trap as everyone else. I couldn't imagine making the decision my rescuer had made—choosing, in a fraction of a second, to risk my own life to save a stranger. The impenetrability of the final outcome made the mental process that could have led there impenetrable as well. It was a problem with no solution, with no hint of which way the solution might even lie, and no matter how many times I turned it over in my mind I couldn't seem to make any progress.

But as it happened, my life had just taken a turn that would begin leading me toward an answer. The previous year I had entered Dartmouth College as one of the many premed students in my class. It was a terrible fit. I quickly found myself fighting to stay awake during the first class period of my first premed biology course. By the second class, I was relying on a classmate's trick of bringing a plastic baggie of Cheerios with me and eating one of them every few seconds for an hour to stay awake. The trick worked, but I knew the bigger picture was futile. What was I doing studying a topic that literally bored me to sleep?

As luck would have it, I had also enrolled in an introductory psychology class that term. From the first class, I was hooked. We covered every question about being a person I had ever thought to ask, and many I hadn't: What is consciousness? How do we see in color? Why do we forget things? What is sexual desire? Where do emotions come from? I can still picture the imposing figure of my professor, the Dumbledore-esque Robert Kleck, striding up and down the aisles of our classroom as he posed questions like, "Is it really true that tall people have better life outcomes?" then pausing dramatically.

Well, do they?? I wondered frantically, all five feet of me. (They do.)

I devoured my textbook, festooning it with highlights and scribbling exclamation points and stars on nearly every page, marking all the insights I wanted to commit to memory. I later marked one page with a sticky note as well. It was a page about teaching sign language to apes, and while I was reading it I had an epiphany: psychology research is something people actually get paid to do—people can do it for a *living*. I decided I was going to become one of those people.

⁂

In 1999, I put this decision into action. I moved to Somerville, Massachusetts, to begin my studies in the doctoral program in social psychology at Harvard University. The program (formerly the Department of Social Relations) is housed in an actual ivory tower—a fifteen-story white stone beacon of behavioral science called William James Hall that rises high above the surrounding buildings in Cambridge. The building is named after the nominal founder of the department, whom many also consider the founder of modern psychology. The students he taught included, among others, Theodore Roosevelt and Gertrude Stein. James was the first of many prominent psychologists to work at Harvard; others included B. F. Skinner and Timothy Leary, the LSD-touting leader of the 1960s counterculture movement. Not all of Harvard's faculty members have been illustrious. Perhaps the most infamous alumnus of the department was Henry Murray, who in the 1950s and 1960s conducted a series of borderline abusive experiments to learn about stress responses during interrogations, which some say were sponsored by the CIA. The subjects in these studies were twenty-two Harvard undergraduates, one of whom, Ted Kaczynski, would go on to become the Unabomber.

That was all in the distant past by the time I began my graduate work under two advisers with no history of abusing any participants whatsoever, Nalini Ambady and Daniel Wegner. Both were among social psychology's greatest minds in their day, for very different reasons.

Among Wegner's many claims to fame was the originality of his ideas. One of these ideas was critical in helping me to understand, if not altruism itself, at least why altruists are so difficult for the rest of us to understand, and why we tend to put them in a psychological box that walls us off from their real experiences and personalities.

With his student Kurt Gray, Wegner explored a phenomenon they called "moral typecasting." The idea is that we automatically and unconsciously divide other people into two categories: moral "agents" and moral "patients." Agents are people who commit moral or immoral deeds—they rescue or rob someone. And patients are the recipients of those deeds—the rescued or the robbed. Agents are the actors, and patients are the acted-upon.

Because moral agents, good or bad, are doers, we focus on their capacity for planning and self-control and ascribe more of these qualities to them. That means we think of rescuers as better planners, as more self-controlled and more capable of complex thought, than the average person. The same goes for robbers. The outcome of their actions may be different, but what heroes and antiheroes share is that both are planners and actors. Typecasting works in reverse too. People who are seen as having a lot of capacity for planning and self-control are also allotted more moral agency. This is why we say that adults have more moral agency than children, and it is one reason why we think it's fair for an adult to be punished more harshly than a child who commits the same crime.

But we don't grant moral agents more of everything. The flip side of having more agency is that moral agents are seen as having *less* of what Gray and Wegner call "experience": emotions like fear and joy, and sensations like pain and hunger. This might be because we see moral agents and patients as non-overlapping categories, and we grant patients all the experience. The patients are the ones to whom something good or bad happened, so we focus on the fear and relief of the rescued child, or the sadness and rage of the robbed shopkeeper. But the feelings of the hero who planned and executed the child's rescue, or those of the criminal who robbed the store, get left by the wayside. This binary worldview of moral agents and patients, recognized as far back as Aristotle, typecasts heroic rescuers as people with great capacities for willpower and control, but little capacity for feelings.

These stereotypes play out nicely in cartoons and action movies, which portray heroes as stoic and impassive. Superheroes like Spider-Man and Batman, or even the nominally human protagonists of the James Bond or *Mission: Impossible* series, might brood sometimes, but we don't get the sense that they feel deep-seated, vulnerable emotions like fear, even when they throw themselves off the edges of buildings or dodge fusillades of bullets. Their job is to skirt death and suffer injuries with barely a grunt. It is nearly impossible to imagine Batman or James Bond screaming in terror.

So do the movies, and our own stereotypes, have it right? Does being an actual, living hero require being resistant to deep, distressing emotions like fear and panic? If we can believe a United States senator, former mayor of Newark, New Jersey, and real-life heroic rescuer, the answer is clearly no.

Back in 2012, when Cory Booker was still the mayor of Newark, he was returning home one evening with two members of his security detail. Approaching Booker's house, they realized that the home of his next-door neighbor was in flames, with smoke pouring through the second-story windows. In the yard outside stood

Booker's neighbor, Jacqueline Williams, screaming that her daughter Zina was trapped on the second story of the house.

Booker said afterward that he acted on instinct. He leapt from the car and raced across the yard past Jacqueline and into the house, with his bodyguard, detective Alex Rodriguez, on his heels. The air inside was thick with smoke. Gasping and choking, Booker and Rodriguez made their way up the stairs to the second-story kitchen. They arrived there to find flames licking up the walls and across the ceiling, accompanied by what sounded like small explosions. That was enough for Rodriguez. His job was to protect the mayor. He grabbed Booker by the belt and started to pull him back out of the kitchen. Booker was having none of it. They tussled for a moment, Booker struggling to free himself. Rodriguez shouted at him, "I can't let you in—that's my job! I have to keep you from danger!"

"*Let me go!* If I don't go in, this lady is going to die!" Booker shouted back.

Reluctantly, the detective released Booker, who plunged further inward.

He couldn't see Zina anywhere. But he could hear her voice calling faintly, "I'm here! I'm here!" from a nearby room.

Booker moved toward Zina's voice through heat so intense and smoke so thick that he could hardly breathe or see. He was disoriented and slowly suffocating. His lungs filled with searing black smoke with each breath. He realized he might be about to die. But he didn't turn around. He groped blindly onward through the heat and smoke until finally his hands found Zina, who was slumped across a bed, limp, and barely conscious. Unless he wanted to have fought his way to her for nothing, his only option was to pick her up and carry her out. So he heaved the forty-seven-year-old woman over his shoulders and staggered back toward the kitchen, which was now engulfed in flames. Burning embers rained down on the exposed skin of his hands as he struggled back down the stairs with the help of Rodriguez. Once outside, both Booker and Rodriguez collapsed onto the ground, the mayor gasping and struggling to

Cory Booker, his burned hand wrapped in bandages, grimaces in response to a reporter's question characterizing him as a "superhero," while Detective Rodriguez looks on. *© NBC Universal*

breathe. EMTs loaded him into an ambulance and rushed him to the hospital, where he was treated for smoke inhalation and second-degree burns to his hand.

Social media exploded with admiration for the mayor. On Twitter, he was painted as a stereotypical stoic superhero, incapable of fear:

> *When Chuck Norris has nightmares, Cory Booker turns on the light & sits with him until he falls back asleep.*
>
> *Cory Booker isn't afraid of the dark. The dark is afraid of Cory Booker.*

Were these accurate reflections of Booker's experience? Hardly. In all of the interviews he gave about the rescue, Booker was blunt about what he had actually been feeling at the time:

"It was a really frightening experience for me."

"When you hear somebody calling for help and you're staring at a room engulfed in flames, it's very, very terrifying. You know, people say bravery, I felt fear."

"Honestly it was terrifying and to look back and see nothing but flames and to look in front of you and see nothing but blackness."

"I did not feel bravery, I felt terror. It was a very scary moment. . . . It looked like I couldn't get back through where I came from."

"When I saw how bad those flames were and felt that heat . . . it was a very, very scary situation."

Frightening. Terror. Terrifying. Fear. Scary. Very, very scary. Booker could not have been clearer. In a terrifying situation, even someone who acts heroically feels terrified. Forget the movies, forget the stereotypes, forget the Twitter plaudits, and fight the urge to typecast: what distinguishes heroes from other people is not how they feel, but what they do—they move toward the source of the terror, rather than away from it, because somebody needs their help.

On the surface this doesn't seem like a big leap in understanding. But as it turns out, it's an enormous one.

2

HEROES AND ANTIHEROES

HEROISM AND ANTIHEROISM both ultimately boil down to suffering. What is heroism except relieving or preventing someone else's suffering? What is villainy except causing it? What this means, unfortunately, is that gaining a better understanding of the roots of goodness and evil, compassion and callousness, requires that somebody suffer. I found this out the hard way—hard as a fist, hard as a slab of concrete—midway through my first year of graduate school, when I was violently assaulted by a stranger. The incident served as a bizarre counterpoint to having been rescued by a stranger. I'm not exactly glad it happened, but it undoubtedly gave me a more complete understanding of the human capacity for callousness and cruelty.

It happened soon after the clock struck midnight on December 31, 1999—that giddy moment when the world realized it would not be ending in a massive global computer meltdown courtesy of the bug known as Y2K. I, along with several of my closest childhood friends from Tacoma, had convened to celebrate the event on the Las Vegas Strip. This was probably unwise. I knew that at the time. The Strip is a bit of a mess even on a quiet off-season night. On the New Year's Eve that marked the dawn of a new millennium, "mess" doesn't begin to describe what it was like. It was chaos, it was Mardi Gras on steroids, it was an endless sea of giddy, drunk, raucous humanity stretching for miles in every direction.

My friends and I were six twenty-three-year-old women who collectively made a second unwise decision, which was that the theme of our night would be "sparkles." Sparkly dresses, sparkly halter tops, sparkly makeup. Also, silly New Year's–themed, glitter-caked cardboard hats and flashing-light sunglasses. We were shooting for glamorous and fell more than a little short. Luckily, Las Vegas standards are not high. When, at the beginning of the night, the six of us and all our sparkles poured out of the elevator and onto the floor of the casino hotel where we were staying, the whole floor burst into spontaneous applause. We heard people shouting, "Whooooo!" and thought we were the most spectacular things in town. It seemed an auspicious start to the evening.

For most of the hours leading up to midnight, we had a ball. Everyone was in a great mood. Televisions in the casinos showed that the clocks had rolled over into 2000 in Australia, and the world had remained on its axis. No computer meltdowns, no shutdowns of city grids. All the people we met, most of whom were roaming around like us in large flocks of twenty-somethings, were ebullient. Buying each other drinks, stopping to pose for group pictures—not something people normally did back then either, when taking pictures required using an actual camera and waiting hours or days to see the results.

But as the evening wore on and our sparkles faded, people's manners started to fade as well. People—men, specifically—started getting grabby. At first it was just the occasional, seemingly errant brush of the hand. But as the hours and drinks piled up, it escalated to grabs and squeezes of breasts and backsides. By midnight, my friends in dresses could feel hands creeping up inside their skirts and down their tops when they stopped to take pictures. I was wearing leather pants and managed to escape some of that indignity, but I lost count of how many times strange men squeezed my ass.

At first, honestly, it was all sort of funny. We were drinking and giddy just like everyone else. It seemed mostly harmless—there were lots of other people around, men and women both, and the Strip was brightly lit and lined with police officers. It never occurred to me that anything worse than a little silly grabbing would happen. Then I saw someone die.

He was young, midtwenties at most. Maybe he was trying to get a better view of the Strip, or impress his friends, or maybe the night's wild frissons just drove him to try something wild. Whatever the reason, he climbed up a metal traffic signal pole on the Strip and ventured out onto the arm that extended over the street. It was impossible to tell from below, but the wires that run through these arms are exposed. His hand made contact with one, and he tumbled, lifeless, to the pavement below. Even if the electric shock hadn't killed him, the fall might have. I read later that he'd landed on his head. That night all I saw was a man up on the pole, and then, a fraction of a second later, he had fallen and the crowd around me was shouting incoherently. The news spread from group to group that the man on the pole was dead. We didn't even know yet if it was true, but the night took on a newly sinister feel.

Getting constantly grabbed rapidly went from funny to tiresome to infuriating. The evening's alcohol was wearing off, and I was tired and my boots were giving me blisters. I remember muttering to myself as I hobbled along, "The next guy that grabs my ass . . ." I didn't even have time to finish the thought before one did. I spun around and glared at him. He grinned proudly back. He was muscular and broad-faced with slicked-back and gelled blond hair. He was also very short. His face, his idiotically leering face, was almost level with my own. I don't know if it was the leer or the gel or just that his grab was the last one I could tolerate, but I slapped him. Pretty hard too.

I saw his grin falter, to be replaced with a flicker of annoyance, and before I had time to think or duck or even turn my head, his fist was hauling back and then smashing into my face with brutal force. The world went wavy and dim as my head snapped back and I crashed down onto the concrete, blood streaming from my broken nose. A murmuring crowd gathered around me. I felt dazed, and I couldn't get the legs and feet around me to come into focus. It took a moment to figure out that the force of the blow had knocked out one of my contact lenses. My friend Heather rushed over. She cradled me as I struggled to gather myself, the blood from my nose trickling down my sparkly top and over her hand.

As she was helping me to my feet, two police officers approached us. They were dragging a man between them—a man whose panicked face I'd never seen. They shook him by the shoulders.

"Is this him?" one shouted. "Is this the guy who hit you?"

His T-shirt was the wrong color. He was too tall. It definitely wasn't him.

"No," I said, shaking my head, "That's not him."

They let him go, and he disappeared into the crowd. I figured my assailant must have done the same. It would be impossible to find him in the swarming sea of people. We turned to leave, and I felt a tap on my shoulder. A woman with blazing eyes stood beside me. Her breath smelled of beer as she leaned in close and murmured in a low and satisfied voice, "I don't know if you saw what happened. A bunch of guys saw that fucker hit you. They chased him down. He's pretty much a smudge on the pavement now."

⁂

The whole incident left me newly tormented. It all seemed so bizarre that I would have been tempted to assume I had dreamed it, were it not for the black eyes blooming across my face the next morning and the fact that my nose was crooked and puffed up to three times its normal size.

I had led a fortunate life in many ways. Intellectually, I knew that violence occurred. My hometown of Tacoma was a hotbed of gang activity throughout the 1980s and 1990s, and the local news was full of shootings and stabbings and muggings. More than one serial killer was picking off Tacoma residents during those years as well. But I had never personally been seriously harmed by anyone. The poet John Keats was correct in observing, "Nothing ever becomes real 'til it is experienced." There really is no substitute for getting your own face smashed in to make you appreciate at a gut level that the world contains people who will actually hurt strangers to serve their own brutal purposes.

My roadside rescuer had made me believe in the possibility of genuine altruism. But more than that, his actions had cast a wider glow over the rest of humanity, whose capacities for altruism were still untested. Perhaps, I'd thought, my rescuer was just one of a vast swath of people who were also capable of great compassion. But what happened to me in Las Vegas didn't stay there. It followed me wherever I went, gnawing at me, whispering in my ear that perhaps I should reconsider my beliefs about human nature. Maybe my rescuer was an anomaly, and my attacker one of many. Who knew how many of the strangers I passed on the streets every day had the capacity to do what he had done? Every man I knew reassured me that under no circumstances would he ever punch a woman in the face, regardless of whether she had slapped him, regardless of how much he'd had to drink. But the fact remained that a whole mass of *other* strange men had rushed my attacker that night and brutally assaulted him next. Did the capacity for such violence lie latent in many or most people? I signed up for a self-defense class, just in case.

My psychology studies offered me no comfort. Here I was at the university my Dartmouth professor Robert Kleck winkingly termed the "center of the intellectual universe," immersed in the best that empirical research had to offer about the nature of human cognition and behavior, and most of it seemed to point to the same terrible conclusions as my Las Vegas encounter. I learned about the infamous case of Kitty Genovese, a Queens, New York, resident who had been brutally murdered on the street outside her apartment building as (so the story then went) thirty-eight witnesses watched silently, none calling for help. The results of follow-up psychology studies by Bibb Latané and John Darley seemed to confirm the reality of the apathetic bystander. I learned about Philip Zimbardo's infamous Stanford Prison Experiment, during which a more or less random sample of Stanford University undergraduates were turned, practically overnight, into cruel and sadistic prison guards simply by donning the requisite role and uniform. So many studies seemed to convey the same message about humans' terrible capacity for cruelty and callousness.

Perhaps the most infamous of these studies—and also possibly the most important—were those conducted by one of Harvard's most eminent PhD students and, later, psychology faculty alumni. Stanley Milgram's research was so controversial it ultimately cost him his Harvard job and tenure. A psychologist of uncanny brilliance and prescience, Milgram is still ranked among the most influential psychologists of the last century (number 46, to be exact). Among his many claims to fame is that he conducted the research that proved "six degrees of separation" is a real thing. In 1963, Harvard hired Milgram away from Yale shortly after he'd concluded another series of studies that may represent the most notorious use ever of electric shocks in psychology research. Like every other psychology major in the world, I had learned as an undergraduate about these studies and the savage cruelty that they showcased.

But also as nearly everyone does, including most psychologists, I initially drew entirely the wrong conclusions from them.

In 1961, Milgram posted newspaper advertisements in New Haven and Bridgeport, Connecticut, inviting local men to volunteer for a study researching how punishment affects learning. When each volunteer arrived in Milgram's Yale laboratory, he was led into a testing room by an angular, stern-looking experimenter in a lab coat. The experimenter introduced the volunteer to a stranger named Mr. Wallace, who, the experimenter explained, had been randomly selected to be the "learner" in the experiment. The volunteer had been selected to be the "teacher." All the volunteer had to do was "teach" Mr. Wallace a long list of word pairs, like "slow-dance" and "rich-boy." Simple enough.

The experimenter showed the volunteer and Mr. Wallace to their seats, which were in adjoining rooms connected by an intercom. Mr. Wallace wouldn't just be sitting, though—he would be tied down. Before the experiment began, and while the volunteer looked on, the experimenter bound both of Mr. Wallace's forearms to the arms of his chair with long leather straps, ostensibly to "reduce movement."

One can only imagine what went through each volunteer's mind at that point. Video footage shows them to be such a wholesome-looking bunch, in their dapper 1960s haircuts and collared shirts. Here they had volunteered for a Yale research study to help science and make a little money, and before they knew what was happening some mad scientist was tying a middle-aged stranger to a chair right in front of them.

The volunteer and experimenter left the room and the experiment began. First the volunteer would read a long list of word pairs through the intercom to Mr. Wallace. Then he would go back to the beginning of the list and read out one word from each pair. Mr. Wallace would try to remember the other word. If he guessed right, they'd move on. If he guessed wrong, he was punished. The volunteer had been instructed to pull one of a long row of levers on a switchboard after each wrong answer. Each lever was marked with a different voltage level, ranging from 15 volts at the low end to a high of 450 volts. Pulling a lever completed a circuit within the switchboard and delivered an electrical shock of that voltage to Mr. Wallace's tied-down arm.

Nearly all the "teachers" went along with the experiment for a while. The experimenter reassured them early on that the shocks were "painful, but not harmful." But as the study progressed and the teacher pulled lever after lever as wrong answers mounted up, the shocks grew stronger. Mr. Wallace started to grunt each time he got a shock, then to cry out in pain. He began complaining that his heart was bothering him. Eventually, the shocks drew long, ragged screams from him, and he bellowed through the wall, "Let me out of here! *Let me out! LET ME OUT!*"

Then he fell silent.

After that point, any teacher who elected to carry on could only grimly continue delivering shocks to Mr. Wallace's unresponsive arm.

Only nobody was expected to carry on that far. Before the study began, Milgram had polled a number of expert psychiatrists about what they predicted would happen. They overwhelmingly agreed that only a tiny fraction of the population—perhaps one-tenth of a percent—would continue administering shocks to a stranger who was complaining about his heart and screaming for mercy.

The experts were overwhelmingly wrong. Fully half of Milgram's volunteers continued administering shocks right through Mr. Wallace's chest pain and screams and well past the point when he fell silent. No external reward motivated their behavior. They would keep their four dollars and fifty cents payment no matter what they did. The only thing urging them along—very mildly—was the experimenter. When a volunteer started to protest or asked that the experiment be stopped while someone checked on Mr. Wallace, the experimenter would reply, "The experiment requires that you continue." Calm prods like this were all it took to induce ordinary American men to subject an innocent stranger to terrible pain, grievous harm, and, as far as they knew, death. One volunteer later said he was so sure that he'd killed Mr. Wallace that he anxiously monitored the local obituaries for some time after the experiment.

Of course Mr. Wallace didn't die—nor did he actually receive any shocks. He wasn't even named Mr. Wallace. He was part of the act, an amiable forty-seven-year-old New Haven accountant named Jim McDonough who had been hired and trained for the role of the study's purported victim.

Nor were the studies aimed at understanding learning. Milgram was really studying obedience to authority—specifically, whether ordinary people would commit acts of cruelty or brutality if told to do so by someone in authority. The research was inspired by the trial of Adolf Eichmann, the Nazi officer who carried out some of the worst atrocities of the Holocaust. Captured in Argentina in 1960 by Israel's Mossad intelligence forces and made to stand trial for his crimes, Eichmann's defense was shocking. He claimed to feel no remorse for his actions, not because he was a heartless monster, but because he had simply been following the orders of those in authority. Later, pleading for his life in a handwritten letter to Israel president Yitzhak Ben-Zvi, Eichmann protested, "There is a need to draw a line

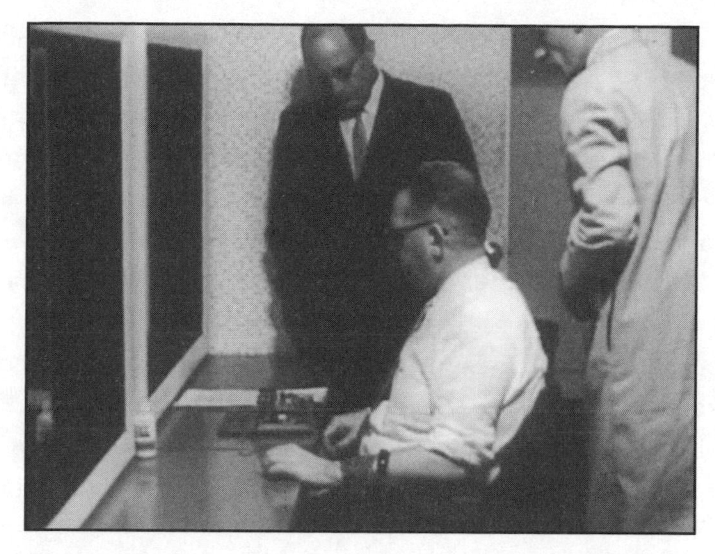

The seated man is Jim McDonough, known to study participants as "Mr. Wallace." They believed they were shocking him during Milgram's studies of obedience to authority. At right is the experimenter, and at left is a volunteer who has just watched the experimenter strap Mr. Wallace's arms to his chair.

Obedience *(1962) (documentary film), filmed at Yale University, released 1965. Courtesy ProQuest.*

between the leaders responsible and the people like me forced to serve as mere instruments in the hands of the leaders . . . I was not a responsible leader, and as such do not feel myself guilty." In essence, Eichmann was claiming that his superiors instructed him that the Final Solution required that he continue.

So he continued.

What Milgram discovered at Yale did not necessarily demonstrate that Eichmann had been telling the truth. Indeed, more recent evidence has suggested that functionaries like Eichmann were not simply cogs in a machine but were proactively and creatively working to advance Nazi causes.

But Milgram did show that Eichmann *could* have been telling the truth. His studies showed that, under the right circumstances, ordinary people will engage in horrific, sadistic crimes if an authority figure who is willing to take responsibility for the outcome instructs them to do so. In another era, under a different regime, Eichmann might indeed have led an ordinary, blameless life. There may have been nothing fundamentally evil about him as a person that would have inexorably led him to perpetuate atrocities. The opposite is also true, of course. Under the right circumstances, ordinary, otherwise blameless people—say, a shopkeeper from Bridgeport, Connecticut—could wind up plotting the torture and deaths of millions of innocent people. After all, the otherwise ordinary, blameless Connecticut shopkeepers and millworkers and teachers in Milgram's studies had, to their knowledge, willingly taken part in torture, false imprisonment, and perhaps even murder in exchange for four dollars and fifty cents.

Milgram would muse in an interview on CBS's *Sixty Minutes* some years later, "I would say, on the basis of having observed a thousand people in the experiment and having my own intuition shaped and informed by these experiments, that if a system of death camps were set up in the United States of the sort we had seen in Nazi Germany, one would be able to find sufficient personnel for those camps in any medium-sized American town."

Nobody disputes Milgram's basic findings, at least not explicitly. But on some level, most people also don't *really* buy them. Deep down, nobody really believes that Adolf Eichmann was an ordinary guy who happened to work for bad managers. Neither do most people believe that a mild-mannered authority figure could induce them personally to override their own moral values and torture someone. Psychology students who watch the videos of Milgram's experiments in classrooms across America every year all reassure themselves, *That would never be me*. So do the Internet surfers who come across the studies on Wikipedia. *That would never be me,* they think. *Maybe some middle-aged, cigarette-smoking, work shirt–wearing, Connecticut-accented, Mad Men–era dupe would be gullible enough to follow those orders, but not me.*

But midcentury conformity has nothing to do with it. Neither does gender or age or social class. Male and female college students in California who were run through a nearly identical experiment only a few years ago acted no better than Milgram's subjects. Versions of the study have been run with varying compositions of study participants across generations and countries—from England to South Africa to Jordan—and they have all replicated Milgram's findings. What do these numbingly familiar results mean? That none of us—not you, not me, not Pope Francis or Bono or Oprah Winfrey or anyone else—can claim with confidence that, if it had been us called into Stanley Milgram's Yale laboratory, we wouldn't have kept pulling those levers too.

The basic findings of these studies are clear and widely accepted. They are also, unfortunately, often misinterpreted. It is easy to draw the conclusion after learning about Milgram's studies or watching the video footage of them that people are uniformly callous and heartless, that within each of us lies a little Eichmann content to inflict terrible suffering on strangers. I certainly did when I first learned about the studies. But in fact, this is not at all what they show.

First of all, when you watch the video footage of the studies, it is obvious that the volunteers were anything but heartless. Even the ones who kept on shocking Mr. Wallace until the bitter end were visibly miserable. They paused and sighed gustily. They buried their heads in their hands, rubbing their foreheads before drying their sweaty palms on their pants. They chewed on their lips. They emitted nervous, mirthless chuckles. Between shocks, they implored the experimenter to let them stop. Milgram reported that at some point *every participant* either questioned the experiment or refused the payment he had been promised. When the experiment finally did stop and it was revealed that Mr. Wallace was only an actor, the participants looked shaky with relief. The major reason the studies are now considered ethically dubious is because of how much the volunteers themselves appeared to suffer.

Second, the volunteers' responses weren't uniform. True, fully half of the volunteers carried out all of the instructed shocks when Mr. Wallace was seated in a separate room from them. But at some point the other half refused to continue. Even more refused in a variation of the study in which all the men sat in the same room. On the other hand, many *fewer* refused when Mr. Wallace was sealed off in a separate room that left him totally inaudible to the volunteers. Milgram ran these and many other permutations of the study designed to make either the experimenter's authority or Mr. Wallace's suffering more or less obvious. The proportion of volunteers who continued carrying out the shocks fluctuated in each permutation, but never did the volunteers behave as a bloc. Inevitably, some continued following the experimenter's orders while others refused—bucking authority to spare a stranger from harm.

It's worth taking a moment to flip things around—to think about what motivated those who ultimately disobeyed the experimenter's orders. After all, why *not* just keep on shocking Mr. Wallace? In theory, if people are uniformly callous, this is what they should all have done. It was the path of least resistance. There was no external reward for stopping. Nor was it likely that the volunteers feared punishment if they kept going—the experimenter repeatedly reassured them that he would take responsibility for Mr. Wallace's fate. Did social norms constrain them? Probably not. In a situation so far out of the ordinary—leather arm straps, lab coats, a shock generator—exactly what social norms would have applied? So if our refuseniks neither anticipated reward nor feared punishment, and weren't trying to adhere to some norm, what was left? What about compassion—simple concern for the welfare of someone who was suffering?

This seems the only likely explanation. The volunteers' entreaties to stop the experiment always invoked Mr. Wallace's welfare. Those who eventually stopped administering shocks said it was because they refused to cause him further suffering.

Even more striking, when you look across all the permutations of the study, it becomes clear that compassion is a *stronger* force than obedience. Think about it this way: When Mr. Wallace was seated in a separate room—invisible and audible only through the intercom—and the experimenter was in the room with the volunteer, the proportion of people who obeyed versus bucked authority was perfectly balanced. Milgram described the influences that the experimenter

Competing fields of force akin to those depicted by Stanley Milgram. An equal balance between forces is achieved only when the stronger force is farther removed from the subject than the weaker force.

Abigail Marsh.

and Mr. Wallace exerted as analogous to fields of force. That an equal balance between these forces was achieved when the experimenter, but not Mr. Wallace, was standing right next to the volunteer suggests that the experimenter's authority was a weaker force than Mr. Wallace's suffering. To exert equal influence, the experimenter needed to be physically closer. When the experimenter and Mr. Wallace were in equal proximity to the volunteer—when both were in the room with the volunteer, or both were outside it—fewer than half the volunteers fully obeyed. The pull of compassion, on average, was stronger than the pull of obedience.

This is an oddly heartwarming message from a study not usually thought of as heartwarming: Milgram actually demonstrated that compassion for a perfect stranger is powerful and common. This is particularly interesting given that Mr. Wallace was hardly the world's most compassion-inducing person. He was a portly, middle-aged man who wasn't especially cute or cuddly-looking and who was a stranger to the volunteers in the study. They'd never met him before, they spoke with him only briefly before the study started, and they were unlikely to ever see him again. He never did anything for them. Why should they have cared about his welfare at all? And yet they did. They ultimately cared more about Mr. Wallace's welfare than they cared about obeying authority, even though their obedience is what everyone remembers.

Now, you could argue that compassion that merely stops someone from zapping a stranger with painful shocks isn't very impressive. More impressive would be compassion that moved volunteers to make some sacrifice to help Mr. Wallace—to give up their payment or undergo some risk to make the shocks stop. Or, even better, to offer to switch roles and receive the shocks in his place. Sadly, Milgram never thought to give his volunteers that chance. But someone else did. Although he is not as well known as Milgram, no social psychologist has uncovered more about the nature of human compassion than Daniel Batson.

Batson holds not one but two doctoral degrees from Princeton University: one in theology and one in psychology. He is linked to Milgram by only one degree of separation: his psychology graduate mentor was John Darley, famed for his studies of bystander apathy. Darley earned his doctoral degree from Harvard in 1965, when Milgram was on the faculty. Darley would probably have taken classes with Milgram, and he certainly crossed paths with him. Darley's student Batson spent his academic career at the University of Kansas conducting research on spirituality, empathy, and altruism—including one study undoubtedly inspired by Milgram's. But Batson's study used electrical shocks to investigate how far compassion would drive ordinary people to *help* a stranger.

Batson recruited his volunteers for the study—all of whom were women—from an introductory psychology course. Each volunteer arriving in the lab was met by an experimenter who told her that the other subject in the study that day was running a little late and could she read a description of the study while they waited? Then the experimenter handed the volunteer a leaflet that described a study that was similar in many ways to Milgram's. It explained that the study was investigating the effects of electric shocks on work performance. As in Milgram's study, Batson's volunteers believed that random chance dictated that the other volunteer would be receiving the shocks instead of themselves. But Batson's volunteers would not be administering any shocks personally. They would merely be watching the other volunteer being shocked via closed-circuit television while they evaluated her performance.

Watching someone get shocked sounds much easier than actually giving someone shocks, and initially it probably was. The other "volunteer," actually an actor posing as an undergraduate, eventually arrived, and the first volunteer watched her on-screen as she introduced herself to the experimenter as Elaine and was escorted into the shock chamber. There the experimenter explained the study to her and attached electrodes, much like those used by Milgram, to her arm. Elaine stopped her at one point to ask how bad the shocks would be. The researcher answered that the shocks would be painful but wouldn't cause any "permanent damage." After this less than reassuring response, the experiment began.

Elaine's job was to remember many long series of numbers. Every so often, while she was in the middle of trying to recite the numbers, the experimenters would administer a strong shock to her arm. It was obvious to the volunteer watching through the monitor how much pain the shocks caused Elaine. With each one, her face contorted and her body jerked visibly. A galvanic skin response reading showed that her hands were sweating profusely. Her reactions grew stronger as the study progressed and the volunteer watched from the other room while trying to evaluate poor Elaine's memory performance. You can imagine her relief when the experimenter eventually paused the experiment to ask Elaine if she was able to go on. Elaine replied that she could, but could they take a break so she could have a drink of water? When the experimenter returned with the glass, Elaine confessed that the experiment was bringing back memories of having been thrown by a horse onto an electric fence when she was a child, a traumatic experience that left her fearful of even mild shocks.

Hearing this, the experimenter protested that Elaine definitely should not continue with the study. Elaine backtracked, saying she knew the experiment was important and she wanted to keep her promise to complete it. The experimenter was briefly stumped. She thought for a moment, then suggested another option. What if Elaine switched places with the volunteer watching from the other room and they carried on with the experiment with the roles reversed?

The experimenter returned to the room where the volunteer sat. She closed the door behind her and explained the situation. The volunteer was completely free to make whatever choice she wanted, the experimenter emphasized: to switch places with Elaine or to continue as the observer. The experimenter even gave some volunteers an easy out—if they decided to continue as the observer, they could just answer a few more questions about Elaine and then they were free to go. They didn't have to watch Elaine any more on the screen. Other volunteers were told that if they chose to continue as the observer, they'd have to watch Elaine get up to eight more shocks.

Put yourself in the volunteer's place for a moment. You've been watching a stranger obviously suffering. Maybe you've been trying to tune out her reactions to the shocks, or maybe you've been thinking of asking the experimenter to stop the experiment. Maybe you've just been feeling relieved that it wasn't you getting shocked. Then suddenly the experimenter appears and turns the tables: it's up to you to decide what happens next. Would you let Elaine keep suffering, or would you be willing to suffer in her place? Would it make a difference if you had to keep on watching her suffer? Batson didn't query any psychiatrists in advance about what they thought the volunteers would do, but perhaps you have a guess. Would any of these teenage women volunteer to receive painful electric shocks to spare a stranger from getting them? How many out of the forty-four volunteers? One or two? Half?

As in Milgram's studies, the researchers varied several features of the experiment, each of which shaped the volunteers' decisions to some degree. One important factor turned out to be how similar the volunteer perceived Elaine to be to herself. Volunteers who perceived Elaine as similar to themselves were twice as likely to help as those who didn't. Whether they would have to watch Elaine continue suffering or whether they could flee also mattered, although less so. But across all the variants of the experiment, a whopping twenty-eight of the forty-four volunteers (a majority by nearly a two-to-one margin) said that they would prefer to take the rest of the shocks themselves rather than watch Elaine suffer through them anymore. Even when offered the chance to escape, over half the volunteers offered to take Elaine's place. In no variation of the experiment was Elaine left high and dry. When asked how many of the remaining study trials they'd be willing to complete in Elaine's place (with the most possible being eight; Elaine herself only completed two), the volunteers in some versions of the study offered to complete, on average, seven.

Stanley Milgram famously revealed that ordinary people are willing to give a stranger painful electric shocks when told to do so by an authority figure. Less famously, he also found that when the power of authority and compassion are pitted equally against each other, compassion ultimately wins. Recall that when the experimenter was in the room with a volunteer and Mr. Wallace was in an adjacent room, half of the volunteers continued shocking Mr. Wallace until the end of the experiment. But when the experimenter's and Mr. Wallace's proximity to a volunteer were equal—when both were in the room with the volunteer, or both outside it—obedience in Milgram's studies dropped below 50 percent, suggesting that the pull of compassion is, on average, stronger than the pull of obedience. Even less famously (sadly!), Batson found that when people are able to choose freely, most will opt to receive electrical shocks themselves rather than let a suffering stranger continue receiving them. Taken together, the real message of these studies is that, when given the opportunity, some people will behave callously or even aggressively toward a suffering stranger—but more people will not. Compassion is powerful. And so is individual variability.

Together, these studies and others like them provided me with my first clues to help solve the mystery of my roadside rescuer. He and all the other drivers who encountered me that night on the freeway found themselves in an identical situation, and it was a situation ideally designed to minimize compassion. I was completely sealed off from the drivers who passed me—trapped inside my car, inaudible and perhaps invisible to them. They wouldn't have known whether I was young or old, similar or dissimilar to themselves, one person or several. The many dangers they would face if they stopped to help would have been all too obvious. Escape from the situation was as easy as not hitting the brakes. Plus, the drivers had only the briefest of moments to decide what to do as they passed by. Under these circumstances, I would never conclude that any of the dozens of people who drove by me without stopping that night was incapable of compassion, any more than I would conclude that about the Milgram volunteers who could neither hear nor see Mr. Wallace and continued shocking him, or the Batson volunteers who left the experiment instead of volunteering to take the shocks for Elaine. The force of my suffering was too far removed from those who passed me to overcome the much stronger and more salient forces of self-preservation and easy escape—for most people. Thank God for variability. Even in these unpropitious circumstances, the drivers did not act as a bloc. One of them stopped to help. And one was all I needed.

Although heroes often argue otherwise, they do seem to be unlike other people in some important ways. Confronted by the same situation—a stranded motorist, a screaming man, a distressed young woman—they are moved to help rather than ignore or flee the situation. Thinking back to Milgram's conception of fields of force, one possibility is that heroes are somehow impervious to the forces that work against heroism, like self-preservation. But this doesn't seem to comport with the experiences of heroes like Cory Booker. He didn't race through a burning building to save his neighbor because he was insensitive to the risks he faced. Far from it—he described himself as having been terrified for his life throughout the ordeal.

Another possibility is that heroes are more strongly affected by the "field of force" that promotes compassion. Perhaps the sight or sound, or even the idea, of someone suffering affects them more strongly than it affects the average person. This seems a critical insight, but unfortunately, neither Milgram's nor Batson's studies give us much information about why or how this might be. Although in some ways Milgram's and Batson's goals were polar opposites (with Milgram interested in forces like obedience that override compassion, and Batson interested in when compassion overrides other forces like self-interest), in other ways their goals were very similar. Both men were trained in social psychology, a discipline that has historically focused on how situations and external events affect people on average rather than focusing on variation among people. Social psychologists ask how events like orders from an authority affect people—on average. How are people influenced by the belief that a stranger is suffering—on average? The advantage of focusing on external events and situations like these is that they can be tweaked. An experimenter can seat a suffering stranger right next to a study volunteer, or choose to make the stranger audible only over an intercom, or to not make him audible at all. Tweak! When these tweaks are coupled with tight control of all the other features of the study—the testing room, the shocks, the experimenter's instructions—you've got a true experiment. And what a true experiment gives you is the most satisfying kind of scientific power—the power to say that the thing you tweaked *caused* the thing you measured. Mr. Wallace's audible cries *caused* more people to stop shocking him. Difficulty of escape *caused* more volunteers to take the shocks on Elaine's behalf.

But the disadvantage of focusing on external events is that you don't get much information about individual variability. The factors responsible for variability often can't be tweaked. They include every biological and environmental force that has ever affected a research volunteer up until the moment he or she arrives in the lab; tweaking factors like these is usually impossible or unethical, or both. What kind of parenting Milgram's volunteers received as children, their IQs, and their personalities could very well have played a role in how they responded. But scientists can't tweak these things. They can't remove babies from their homes and foist them on new parents to study parenting, or inflict IQ-altering brain damage, or give people personality-altering drugs. Experiments like these would be as horrific as Eichmann's atrocities.

So we have to do the best we can without tweaking these fundamental features. We observe and measure parenting and intelligence and personality, try to control for extraneous variables, then statistically map possible causes onto possible effects, all with the knowledge that we may still be missing something. For instance, if an aggressive adult had harsh and punitive parents, perhaps the harsh parenting caused his aggression. Or perhaps not. People also share genes in common with their parents. So another possibility—one of many—is that observed correspondences between harsh parents and aggressive offspring result from common genetic factors causing them all to act out.

The Milgram volunteers who continued shocking Mr. Wallace all the way down the switchboard intentionally and knowingly caused him harm, which is the definition of aggression. Let's say that the volunteers who tended to do this were raised by harsher than average parents. Even if this were true, we couldn't say that harsh parenting caused the volunteers to act more aggressively because there are too many other possible alternatives. It comes back to the old trope about correlation not implying causation. Keep this in mind whenever you hear about developmental studies linking some behavior in parents—harsh discipline or breastfeeding or using complex language or anything else—to some outcome in children. Just because one event precedes another does not mean that it caused the other. Many of these studies aren't designed to tease apart the roles of genetic and environmental factors, so they can't clearly establish cause and effect.

Luckily, there are ways of getting around problems like this. One is to take advantage of *natural experiments*. Natural experiments result when a variable is tweaked by someone, or something, other than a scientist. They are rarely natural and are not true experiments, but they are invaluable nonetheless. One well-known example is adoption studies. It's clearly unethical for scientists to take babies from their biological parents and give them to unrelated adults to raise, but it's fine—admirable even—for adoption agencies to do the same thing. And what's not unethical is for scientists to *study* children who have been adopted to untangle the effects of genes and parenting. Adoptions "naturally" disentangle genes and parenting by ensuring that one set of parents contributes genes and an unrelated set contributes only parenting. So scientists can study adopted children to learn about how genes and parenting contribute to nearly any outcome in children.

Another way to disentangle genetic and environmental effects is through studies of twins. Because identical twins share 100 percent of their genes whereas fraternal twins share on average only 50 percent (just as any other biological siblings do), the contributions of genes and the environment can be teased apart by studying similarities and differences between identical and fraternal twins. An even more powerful approach is to combine these methods and study identical and fraternal twins raised by their biological or adopted parents. As a result of such studies we know, for example, that identical twins are very similar to one another across multiple physical and psychological indices, even when they are raised in different households. In fact, when raised apart they are sometimes more similar to one another—in terms of their IQs, say—than fraternal twins raised in the same household. A study like this provides compelling evidence for genetic contributions to intelligence.

The results of twin and adoption studies show that some outcomes are almost entirely inherited. To no one's surprise, for example, adopted children's eye color corresponds more strongly to that of their biological parents than their adopted parents. The heritability of human eye color—meaning how much of the variation in eye color results from inherited factors rather than environmental factors—is about 98 percent. Environmental factors contribute almost nothing.

This is how researchers could determine with certainty the eye color (blue) of the English king Richard III, who died more than 500 years ago, using DNA samples extracted from his bones. The heritability of other physical features also tends to be high. Height is about 80 percent heritable, meaning that most variation in height results from genes. The other 20 percent largely reflects the effect of nutrition or illness. At least, this is how it works in typical modern, prosperous societies.

A caveat is that the heritability of some traits, like height, may fluctuate depending on the environment. For example, when food is scarce, the heritability of height decreases. This is because genes encode people's maximum *potential* height—the height they can achieve with adequate health and nutrition during childhood. Food scarcity prevents people from reaching their potential, and the greater the deprivation, the greater the difference between their genetic potential and their actual adult height. In a malnourished population, those children who get 70 percent of the calories they need end up even shorter than the children who get 90 percent of what they need. As a result, widespread environmental factors like food availability account for much more than 20 percent of the differences in malnourished children's heights. And as the proportion of the variability accounted for by the environment goes up, the proportion accounted for by genes inevitably goes down.

When children are getting enough food, however, more of it will make no difference. Children who consume 100 percent of the calories they need—enough to compensate for the calories they expend through activity and growth—will not be shorter than children who consume 110 percent of the calories they need. Once you have enough food to reach your maximum potential height, getting more food has no further effects. This is why public health efforts aimed at improving a population's well-being tend to focus on reducing poverty rather than increasing wealth: even small reductions in poverty can improve overall outcomes in a way that increases in wealth never will.

Even in prosperous populations, the heritability of other physical traits is somewhat lower than it is for height. For body weight, heritability hovers around 50 percent. This makes sense, in part because body weight has no maximum potential value, nor is it fixed by adulthood. So your parents' choices and other environmental factors that shape your diet and lifestyle can more strongly shape your body composition than your height. But that 50 percent genetic contribution should not be ignored—body shape is not infinitely modifiable. No biological offspring of stocky parents will ever be rail-thin, even if the child is adopted and raised by paleo-vegan Pilates devotees. No diet could turn Kim Kardashian into Kendall Jenner, her lanky half-sister. Kendall carries the genes of tall, lanky Caitlyn Jenner, whereas Kim's father was the short, stocky Robert Kardashian Sr. Biology is not destiny, but it does place limits on destiny.

This is true for essentially every complex human trait, from physical traits like body shape or facial appearance to psychological traits like aggressiveness or extraversion. According to the famed behavioral geneticist Eric Turkheimer, the first law of behavioral genetics is that all human behavioral traits are heritable. Like body composition, most psychological traits—our mental composition, you might say—are about 50 percent heritable. A massive study reported in the journal *Nature Genetics* showed that, across fifty years of studies of hundreds of thousands of pairs of twins, genes account for, on average, 47 percent of variance in cognitive traits like intelligence and memory and 46 percent of variance in psychiatric traits, including aggression. Parenting and other environmental factors undoubtedly shape outcomes as well, but genetic factors are at least as influential—and often more influential. This helps explain why twins adopted into separate families and later reunited find themselves tickled by the similarities they discover—from hair color to preferred hairstyle to preferred hobbies—despite their separate upbringings.

In Milgram's era, most psychologists would have considered delving into the heritability of aggression, or any other personality variable in humans, a fool's errand. From the early twentieth century and extending well into the 1960s, the tenets of a school of thought called behaviorism dominated psychology. Behaviorists like John Watson and B. F. Skinner viewed observable variations in animal and human behavior as primarily a result of their learning histories. If an organism—a pigeon, a rat, a monkey—had been previously rewarded (or "positively reinforced," as the behaviorists termed it) for pushing a button, it would come to push the button more. If it had been punished for pushing the button, it would push it less. Two animals in adjacent cages that pushed their buttons different numbers of times must have experienced different prior outcomes for doing so. The behaviorists' views were very influential—Skinner (yet another of the Harvard Psychology Department's famous faculty) is today considered the single most influential psychologist of the last century.

Skinner's experiments were beautifully designed and their results compelling. The ingenious "Skinner boxes" that he created to test his predictions were preserved in a lovingly curated exhibition in the basement of William James Hall that I used to pass on my way to my classes there. I marveled at the ingenuity of the little boxes festooned with elaborate arrays of tiny wires and pulleys and buttons and drawers. Of course, Skinner's experiments used little metal boxes because all of his participants were rats and pigeons. In truth, the scope of his research was very narrow. He measured only simple behaviors that could be tested in one of his boxes, like lever-pulling and button-pecking; then, like other behaviorists, he extrapolated wildly from his findings, arguing that all variability in all animals' behaviors—from rat aggression to human language and love—was best understood as resulting from learning histories. Skinner famously mused in his novel *Walden Two*, "What is love except another name for the use of positive reinforcement? Or vice versa."

Likewise, the thinking went, two children in adjacent houses with different levels of aggression must simply have received differential reinforcement for aggression along the way—one rewarded for aggressive behaviors more than the other. The rewards in question needn't be cookies or stickers. Only the world's worst parents would reward aggression with literal prizes. But all sorts of other inadvertent behaviors on the part of parents could reward aggression, in theory. When aggression begets attention, even in the form of yelling or criticism, it can be more rewarding than no attention. A child who is ignored most of the time except when he hits his brother—in which case he gets yelled at—might actually prefer the yelling. Or if hitting his brother gets him something else he wants—his brother out of his bedroom, a toy his brother was holding—that's a reward too. According to Skinner, if we could perfectly control the rewards and punishments that children receive from their rearing environments, we could eliminate undesirable behaviors like aggression entirely.

But there is absolutely no evidence that this is true. Rewards clearly do influence behavior, as do punishments. But heritability studies prove without a doubt that they are not the only influences. The heritability of aggression is consistently found to be around 50 percent, and for some forms of aggression it is as high as 75 percent. If that much of the variability in children's aggression can be predicted from genetic differences among them, genes must play a major role in promoting aggression.

What all of this means is that understanding aggression—and the compassion that can inhibit it—requires more than looking at people's behaviors inside a laboratory, where various environmental variables can be tweaked to make people behave more or less compassionately. A complete understanding of the roots of aggression and compassion also requires looking at deeply rooted, inherited variables that affect how compassionately people behave—that produce the variation in how people respond to various tweaks. Perhaps the most infamous and compelling such variable is psychopathy.

⁎

Psychopathy (pronounced *sigh-COP-a-thee*) is a disorder that robs the human brain of the capacity for compassion. It is characterized by a combination of callousness, poor behavioral control, and antisocial behaviors like conning and manipulation. Psychopaths need not be violent, but they often are. Only about 1 or 2 percent of the American population could be classified as true psychopaths, but among violent criminals the number may be as high as 50 percent. Psychopaths are marked by their tendency to engage in *proactive aggression*—acts of violence and aggression that are deliberate and purposeful rather than hot-tempered and impulsive.

Psychopathy is also highly influenced by genes, with a heritability quotient that may be as high as 70 percent. That this surprises many people I encounter reflects a common view of human aggression that is often, whether they know it or not, colored by the long shadow of behaviorism. Most people assume that violent, callous individuals must be the outcome of highly abusive or neglectful homes. But this simply isn't true.

Take Gary Ridgway, the middle son of Mary Rita Steinman and Thomas Newton Ridgway. Gary and his brothers were raised in McMicken Heights, Washington, just north of where I grew up in Tacoma. The family was poor, no doubt: Thomas drove trucks off and on for money, and the family was crammed into a 600-square-foot house. Mary was a bossy and dominating mother—a "strong woman," as her oldest son Greg later recalled. She and her husband had fights that sometimes turned violent—she once broke a plate across his head during a family dinner. But Gary also remembered her as a kindly figure who did jigsaw puzzles with him when he was a small boy and helped him with his reading. There was no sign of true abuse or household dynamics outside the range of normalcy for a family in the 1960s. And Gary's brothers grew up to lead ordinary lives.

Not so for Gary, who grew up to become the Green River Killer, the most prolific serial murderer in American history. He is now serving a life sentence for forty-nine confirmed murders, and he has claimed that he committed dozens more. His first attempt at murder took place around 1963—the same year, as it happens, that Milgram was conducting his studies on obedience.

Gary was about fourteen years old and on his way to a school dance. Walking through a wooded lot, he ran across a six-year-old boy. Almost without thinking, he pulled the boy into the bushes and, using a knife he always carried with him, stabbed the boy in the rib-cage, piercing his kidney. He quickly withdrew the knife and watched blood gush from the wound. Then Gary walked away, leaving the boy to die—or live. He wasn't particularly concerned either way, except that he hoped that if the boy lived, he wouldn't be able to identify him. (The boy did live, but never identified Ridgway as his assailant.) Later on, Ridgway couldn't even pinpoint why he had done it. It had felt like it just happened, much as other bad things often seemed to just "happen" for him—gleaming rows of windows shattered by rocks, birds felled by a BB gun, a cat suffocated in a picnic cooler.

As Ridgway neared adulthood, things turned much darker. Relentless sexual urges awakened within him. Combined with the callousness and delight in the power of killing he already possessed, those urges turned Ridgway into an insatiable sexual sadist who raped and murdered at least forty-nine girls and women, most of them teenage runaways and adult sex workers around the town of SeaTac in the 1980s, while I was in elementary school some thirty miles to the south.

Ridgway was an unusually depraved personality even compared to other murderers—"a lean, mean killing machine," as he called himself. Mary Ellen O'Toole, a famed FBI profiler and expert on psychopaths, spent many hours interviewing Ridgway, and she has told me that he is one of the most extreme, predatory psychopaths she has ever encountered.

Ridgway lured his victims into trusting him by showing them pictures of his young son Matthew, or leaving Matthew's toys across the seat of his truck. After kidnapping them, he assaulted and killed the women and girls in ways that were often gruesome or bizarre, even by the standards of a culture inured to the horrors of *CSI* and *True Detective*. Most of his victims were suffocated or strangled, and all of them showed signs of sexual assault. Their arms and hands bore bruises and other injuries. Oddly shaped stones were sometimes found in their vaginas. Several of their bodies were festooned with branches or loose brush. One victim, a twenty-one-year-old named Carol Christensen, was found lying in the woods with a paper bag over her head, twine wrapped around her neck, and a wine bottle lying on her stomach. A trout was draped across her neck, and another lay on her shoulder.

People are hungry for details about psychopaths. I have learned that if I want to start an hour-long conversation with a stranger, I need only mention that I study psychopathy. (If I want to be left alone, I say I'm a psychology professor, which sends people running for the hills.) At least ten books have been written about Ridgway, including one by his defense attorney, Tony Savage, and another by Ann Rule, the queen of true crime. Why the fascination? I don't fully understand it myself, but I think it is partly because psychopaths, especially the really ghastly ones like Ridgway, are simultaneously so terrifying and so hard to identify. Even psychopaths who commit strings of unimaginably awful serial murders are often shockingly normal on the surface. And not so-normal-they-seem-creepy normal. Actually normal. Wave-to-their-neighbors-on-the-way-to-work normal.

Tony Savage emphasized this in a 2004 interview with Larry King. "Larry," he said, "I keep telling people, you could sit down and talk with this guy at a tavern and have a beer with him, and twenty minutes later, I'd come up and say, 'Hey, this is the Green River monster,' and you would say, 'No way!'" If you think about it, this has to be true. If psychopaths were obviously creepy or "off," they couldn't commit long series of crimes. They wouldn't be able to convince their victims to trust them or to evade detection for long.

Their seeming normalcy distinguishes psychopaths from murderers who are *psychotic*—a common confusion, but an important distinction. Psychosis is the inability to distinguish fantasy from reality. It is a common symptom of schizophrenia and bipolar disorder, and usually takes the form of delusional beliefs or hallucinations. People who are psychotic might believe that they are being followed by the CIA or sent secret messages through billboards or their televisions, and they might hear voices telling them to do terrible things, including, sometimes, to commit acts of violence. (Most people who are psychotic are not violent. But the results can be devastating when they are, sometimes because they are both psychotic and psychopathic—a truly awful combination.) Recent mass killers like Jared Loughner, who shot former congresswoman Gabrielle Giffords and eighteen others in a Tucson, Arizona, parking lot, and James Holmes, who shot eighty-two people in an Aurora, Colorado, movie theater, were psychotic. People who knew them found them odd and alarming, and even in photographs it is easy to see how disturbed they were. But mass shooters like Loughner and Holmes don't need to convince anyone to trust them or evade detection, because they commit their crimes all at once and out in the open and often intend to die anyhow, either by self-inflicted or police-inflicted wounds.

As scary as mass killers are, serial killers are somehow scarier, perhaps because the most frightening kind of danger is the kind that cannot be predicted in advance. Not all serial killers are psychopaths, but a lot of them are. And if psychopaths genuinely come across as normal, there is no easy way to steer clear of them, making them that much more frightening. My guess is that the pervasive fascination with psychopathy in part reflects a desire for details that will somehow give psychopaths away—nonverbal "tells" like unusual patterns of eye contact, or signature biographical details like childhood bed-wetting or fire-setting. Maybe people think that if we can just find the clues that mark people as psychopaths, we can avoid them or round them up and lock them away. This could be why the myth that psychopaths result from abusive upbringings is so persistent. It seems plausible, it is sometimes true (Ted Bundy and Tommy Lynn Sells are two notorious psychopathic murderers who experienced terrible abuse as children), and it might be the kind of signature detail we could use to isolate the budding psychopaths among us.

Some of Ridgway's biographers have fallen prey to just this temptation—trying to link his gruesome career as a mass murderer to his parents' fighting, or the way his mother bathed him. But it's just not that simple. Thousands of children witness their parents fighting, sometimes violently, every year. Many thousands more, sadly, are abused or neglected, sometimes horribly so. But (thankfully) we don't have thousands of serial murderers running around in the aftermath of this mistreatment. If childhood mistreatment alone caused people to become psychopathic killers on the scale of Gary Ridgway, our society would make a zombie apocalypse look like Disneyland.

Without a doubt, the maltreatment of children is a terrible thing. Children who are abused or neglected or witness violence frequently experience all kinds of negative outcomes later in life. They often develop, not surprisingly, exaggerated sensitivity to potential threats or mistreatment, and they sometimes overreact aggressively to it. This is called *reactive aggression*—angry, hotheaded, impulsive aggression in response to being frustrated or provoked or threatened. If your significant other threatens to leave you and you throw your glass at him, this is reactive aggression. If someone bumps into you on the sidewalk and you turn around and shove him, this is reactive aggression. If a strange woman slaps you after you grab her ass and in response you haul back and punch her in the face—again, reactive aggression. This kind of aggression is relatively common, and it often crops up in people who are depressed or anxious or have experienced serious trauma.

But this is *not* the primary problem with psychopaths. Psychopaths can be quite impulsive and do often engage in reactive aggression, but recall that what really sets them apart is *proactive* aggression—the cool-headed, goal-directed kind of aggression, the seeking-out-vulnerable-women-to-rape-and-murder kind. Child abuse and neglect *don't* seem to promote this kind of aggression. There is almost no evidence of any direct, causal links between parental maltreatment and the proactive aggression that sets psychopaths apart. It's not like people haven't looked for evidence, but well-controlled studies just don't find it.

For example, one study conducted by psycopathy expert Adrian Raine and his colleagues at the University of Southern California looked at reactive and proactive aggression in more than 600 ethnically and socioeconomically diverse pairs of twins in the greater Los Angeles area. They tracked the twins over the course of their adolescence, which is the time when aggression tends to become most pronounced. The researchers found that genetic influences contributed about 50 percent to persistent reactive aggression across adolescence, with the rest resulting from environmental influences. But genes contributed a whopping 85 percent to persistent proactive aggression. And none of the remaining 15 percent was attributable to what are called *shared environmental influences*, which include any influences that affect children within a family similarly, like poverty, the type of house or neighborhood they live in, or having parents who fight or are neglectful. These shared influence variables—*even all added together*—don't seem to predict the course of proactive aggression in adolescents.

This of course leaves an urgent, open question: what *does* cause psychopathy? Through a series of very fortunate events, I got the opportunity to take part in seeking out the answers.

⁎

In 2004, I was nearing the completion of my PhD and finishing up my doctoral dissertation. What I needed next was a job. I knew I wanted to continue in academic research, but at twenty-seven, I wasn't ready to begin a professorship. I had managed to secure a tenure-track offer from a small, selective, rural college, but I couldn't bring myself to accept it. The school was too small and too rural, and I didn't feel ready to face the slog that assistant professors face on their way to tenure. The obvious alternative was a postdoctoral fellowship, which is sort of the equivalent of a medical residency for PhDs. Postdoctoral fellowships provide doctoral graduates with a few extra years of training in the laboratory of an established investigator. Postdocs are a terrific way to acquire training in new research techniques and publish original research before tackling a professorship.

I started looking for a postdoctoral position based mostly on geography. I was engaged to be married, and my fiancé, Jeremy, whom I had started dating at Dartmouth, was a US Marine who had nearly completed his four years of service. Of all the cities in the country, the one with the most and best professional opportunities for a former Marine with a Dartmouth degree in government is Washington, DC. So I started looking there. The Washington area contains several major research universities, and better yet, the National Institutes of Health is in Bethesda, Maryland, just a few miles outside the District.

As an aside, of all American city names, "Bethesda" may be my favorite. It is named for Jerusalem's Pool of Bethesda, the waters of which are described in the biblical Gospel of John as possessing extraordinary powers to heal. The Bethesda in Maryland may be less poetic, but it also possesses extraordinary healing powers. The NIH is far and away the biggest supporter of medical research in the world. The billions of dollars in grant funding it has awarded to researchers around the world over the last several decades have underwritten discoveries of treatments for diseases ranging from cancer to HIV to schizophrenia that have healed countless suffering people.

The NIH also supports a smaller number of scientists—about 6,000—who conduct research on its Bethesda campus. The "intramural researchers," they are called. The intramural resources at NIH are abundant, and its location right outside Washington, DC, made it a perfect spot for me geographically. But what were the odds that I could find a position there? Most NIH researchers do medical research and have degrees in medicine or biology or chemistry. Even at the institute where most psychology and neuroscience research is conducted, the National Institute of Mental Health (NIMH), the researchers are mostly psychiatrists and clinical psychologists. Was there any place there for a social psychologist casting around for a postdoc?

I sought help from a former graduate school colleague, Thalia Wheatley, who was also a social psychologist and had recently started a postdoc at NIMH. Did she know of any researchers on campus who might have a postdoc position for me? She suggested a few names, the last of which was James Blair. "Oh, he'd be perfect for you!" she said. "You're interested in empathy, and he studies psychopaths."

"James Blair?" I repeated. "Wait, that's not *R. J. R. Blair,* is it?"

R. J. R. Blair (alternately R. Blair, J. Blair, R. J. Blair, or J. R. Blair) was the researcher with the hard-to-pin-down initials who I knew to be among the world's foremost researchers on the neural basis of psychopathy. I was very familiar with his work, having cited seven of his research papers in my dissertation, but the bylines on those papers said he was at University College London. His relocation to the NIMH was so recent that there was no scholarly record of it. Thalia laughed. "Yes, R. J. R. Blair is James Blair. And I think he is looking for a new postdoc. I have a meeting with him next week and I can find out."

I was ecstatic. Thalia was right, this was perfect. This was better than perfect.

Although I was earning my degree in social psychology, where the focus is historically on how people as a whole respond to external influences, over the course of my graduate studies I'd been moving in the direction of studying differences among people. As I sought predictors of altruistic responding in my laboratory studies, I noticed that the individual differences were often more important than the laboratory manipulations that were my initial focus.

For example, one of my dissertation studies aimed to replicate an altruism paradigm developed by Daniel Batson. Batson's primary focus was on the relationship between empathy and altruism. I should note that Batson used the term *empathy* to mean what most researchers now refer to as *empathic concern* or *compassion* or *sympathy*—namely, caring about others' welfare. The term *empathy* is more commonly used to mean simple apprehension of another's emotional state, or sometimes sharing that state. If you look frightened and I correctly detect how you are feeling and show physiological changes like an increased heart rate or sweating hands, or if I report feeling upset myself, we can say I've experienced empathy. If I also express the desire to *alleviate* your distress, that's empathic concern or compassion. The processes are related but distinct.

Batson manipulated empathic concern by asking some volunteers to focus on the thoughts and feelings of a woman named Katie Banks, whose sad radio interview they were listening to. In it, Katie described the terrible hardships she was experiencing following the deaths of her parents, which had left her to care for her young siblings while trying to complete college. Other volunteers were asked to focus on the technical details of the broadcast. Batson reliably found that instructing volunteers to focus on Katie's feelings caused them to offer more help to Katie afterward. I found this in my own research too. Volunteers listened to a similar radio interview and afterward were given the opportunity to pledge money or time volunteering to help Katie. (In my study, Katie was actually me putting my college theater training to good use while reading from the same transcript Batson had used.) The research assistants running the study gave the volunteers envelopes in which to seal their pledges so that their decisions would remain anonymous. Like Batson, we found that the volunteers instructed to focus on Katie's feelings experienced more empathic concern and pledged more time volunteering to help her than did those asked to consider technical details of the broadcast.

But this manipulation was not the only, or the best, predictor of how much time people pledged. After the volunteers had listened to the broadcast, we gave them other forms to fill out and tests to complete. One of them was a test of facial expression recognition. Volunteers viewed twenty-four standardized photos of young adults posing expressions of anger, fear, happiness, and sadness and tried to identify each expression using a multiple-choice format. Some of the expressions were obvious, but others were subtle. One dark-haired woman's fearful expression was betrayed by only the faintest elevation of her upper eyelids and slightly parted lips.

After the study, my research assistants and I tallied up the volunteers' accuracy in recognizing each of the various emotions and plotted them against their donations to Katie. What we found surprised me a little. Volunteers' ability to recognize happy expressions was actually a negative predictor of donations: the volunteers who pledged the most time to help Katie were worse than average at recognizing happiness. But the most generous volunteers were better than average at recognizing fearful facial expressions. Even more surprisingly, the power of fear recognition to predict pledges was statistically stronger than the effect of the empathy manipulation. When it came to predicting pledges of money to help Katie, the empathy manipulation didn't predict anything. Instead, the most powerful predictor of donations of money to Katie, by a mile, was individual variation in the ability to recognize others' fear.

I followed up this puzzling finding with more studies, which kept showing the same thing: the most reliable predictor of altruism, across different tests and groups of participants, was how well people could recognize fearful facial expressions. This was a better predictor than the ability to recognize any other facial expression, and it was a better predictor than other traits that are sometimes touted as promoting altruism, like gender, mood, and how empathic people report themselves to be. It was a weird result. I knew it at the time. It later went on to be selected by the psychologists Simon Moss and Samuel Wilson as one of the "most unintuitive" psychology findings of 2007. It wasn't an anomaly, though. Subsequent research has also linked sensitivity to fearful expressions to altruism and compassion in both adults and children across different cultures.

There was one set of data out there that could make sense of these findings. But it wasn't data collected by a social psychologist—it had been published by none other than James Blair. And, lucky me, he did offer me that postdoc position. That meant I'd soon be working alongside him in his new NIMH lab, digging deeper into the brain basis of the capacity to care for others by conducting the first-ever brain imaging research on psychopathic teenagers.

3

THE PSYCHOPATHIC BRAIN

ON THE THIRTIETH of March, 2004, Jeremy and I drove away from Somerville, Massachusetts, bound for Washington, DC, with my cat and a U-Haul full of rickety furniture. Two days later, on April 1, I arrived at the NIH to start my new job.

Upon arrival, my initial impression was—April Fool!—that a colossal prank had been played on me and I actually didn't have a job there. Coming through the gates for the first time, I tried to locate my building on the campus map. The NIH campus holds, depending on how you count, a jumble of about eighty haphazardly numbered buildings—8 is across from 50, which abuts 12. After several minutes of scanning the map, I was forced to concede that my building number wasn't on it. I asked the security guards for help, but none of them had ever heard of it. "Fifteen Kay? What is that? Is that an NIH building?"

In desperation, I started wandering through the vast, rolling campus, and miraculously I finally stumbled upon my destination—a sweet Tudor-style cottage not at all befitting the sterile designation "15K," sitting tucked away on a daffodil-strewn hillside in a remote corner of the campus. It was so small that NIH maps rarely bothered labeling it. No one inside seemed to know who I was or why I had come. A secretary asked me for any paperwork that could confirm

I had been hired, and I realized that no one had ever sent me any. I tried to locate James, but none of the office doors had names or numbers on them. When I finally found his office, it was dark and shuttered. "What the hell is going on? What kind of place is this?!" I wondered in a fury.

As I would come to learn, the NIMH is a byzantine kind of place, especially for newcomers. Paperwork confirming my postdoc was probably wending its way through a labyrinth of cubicles somewhere on campus, but it hadn't yet gotten a final stamp of approval. No matter. I had a computer, I had a desk, I eventually located James, and I was ready to get to work.

My first goal was to publish my dissertation. In a series of five studies, I had found that sensitivity to fearful facial expressions is a reliable predictor of empathic concern in response to others' distress. The first study found that participants who were best at recognizing fearful expressions offered more money and time to help Katie Banks, the distressed woman I had portrayed in the radio interview. The remaining studies found that volunteers who were best at recognizing fear also evaluated strangers' physical appearances more kindly if they thought those strangers would receive the evaluations, and they expressed a greater desire to help distressed strangers described in short vignettes.

James Blair's research could be the key to understanding these odd findings. In 1995, James had published a novel hypothesis of psychopathy. A hallmark of psychopaths is their frequent perpetration of proactive aggression—cold, purposeful aggression, be it physical, verbal, or social—aimed at achieving some goal. Criminals who kill people after robbing them so they can't be identified later, or who use threats of violence to extort money, are often psychopaths. James proposed that the mechanism that prevents most of us from engaging in these sorts of behaviors—and that appears to be malfunctioning in psychopaths—is a system he termed the Violence Inhibition Mechanism, or VIM (later updated and renamed the Integrated Emotion Systems model).

James developed the idea of the VIM based in part on the work of animal behavior experts like Konrad Lorenz and Irenäus Eibl-Eibesfeldt, who observed that, among group-living animals in the wild, conflicts over resources or status can be quelled before actual aggression erupts through the use of body postures and vocalizations that send specific signals. Take wolves, which are good analogs to humans for two reasons. First, the organization of their packs is not unlike small bands of prehistoric humans or modern hunter-gatherers. Both consist of smallish, mutually dependent, and interrelated groups of adults and their young who work cooperatively to defend their territory, care for juveniles, and hunt for food. Second, many behaviors that wolves use to communicate are familiar to us because they have been retained in wolves' domesticated descendants—our dogs.

If, during a walk in the woods, you were to encounter a wolf that approached you with its fur standing on end, its body stiff, and its tail and head held high, growling in a low tone, you would need no translator to tell you how much trouble you were in. The wolf would be telling you that loud and clear. Not because it saw you as prey, mind you—wolves don't bother communicating like this with animals they are planning to eat. This wolf's behavior is an intimidation display—one that is meant to be seen and that signals it views you as a competitor or a threat, perhaps because you wandered too near a kill or its pups. So what would you do?

If you were you (a human), you'd be in a bad spot. Wolves are usually fearful of humans, but when they're not, there is little that a lone, unarmed human can do to fend them off. Outrunning a wolf is impossible, as is, probably, winning a physical fight with a creature whose jaws can crush a moose femur. Your best bet would be to walk backward slowly, avoid eye contact, and pray.

If you were another wolf, however, you'd have much better options. One would be to exploit the Violence Inhibition Mechanism. Through it, you might be able to make the approaching wolf not *want* to attack you anymore. You'd need to start by avoiding eye contact and crouching down low. No, even lower. Your goal would be to try to look a fraction of your actual size. Even better, you might roll on your back, fold your legs in close to your body, flatten your ears, and emit a few whimpers—the higher-pitched and more helpless sounding the better. If you really wanted to go for the gold, and if the wolf came close enough, you could try licking the bristly underside of its jaws or peeing on yourself. If you have owned a very submissive dog before, you have probably seen these behaviors play out. If not, it all might seem distasteful and counterintuitive to you. Why signal how weak and pitiful you are to an animal that is threatening you?

Of course, you wouldn't do any of these things if you were threatened by a nonsocial creature like a rattlesnake or a shark. Go ahead and try licking a rattlesnake and see where it gets you. But the social wolf is acutely attuned to signs that another member of its species is raising the equivalent of a white flag. In adopting postures and vocalizations that make itself look smaller and weaker, a wolf under attack can signal that it will not—*cannot*—contest its would-be attacker's dominance and power, rendering an actual physical battle unnecessary. Weak and subordinate wolves display these cues all the time during conflicts with stronger, more dominant wolves, who largely inhibit their aggressive attacks in response. Such a system makes pack life better for everyone. Because they use body language and vocalizations to communicate their relative power and aggressive or non-aggressive intentions so effectively, wild wolves rarely resort to actual violence.

We humans may not pee on ourselves or roll onto our backs to signal fear and submission to one another, but we do have other cues that serve similar purposes. Body postures, vocal cues, and facial expressions that communicate dominance and subordination are built into our communicative repertoires just as surely as they are in wolves. And as is true for wolves, fearful and submissive cues tend to make a person appear physically smaller and weaker. Fearful body movements are cringing and crouching—think a hunched posture with shoulders huddled and hands and arms drawn in close to the body to shrink the visual silhouette. Fearful vocalizations are high-pitched, like the cry of a small creature with a small and minimally resonant larynx. And fearful facial expressions convey vulnerability and powerlessness through a combination of round, widened eyes, raised and upwardly angled brows, and a grimace. These cues are designed to appease—to literally disarm a potential attacker. Imagine hitting someone who was whimpering and cringing and looking terrified in front of you. It's hard to contemplate without feeling like a terrible person. Consider your potential for Violence Inhibited.

James laid out how the Violence Inhibition Mechanism causes typically developing children to acquire an aversion to hurting others. Young children are almost always at least a little aggressive, with age two being the most violent year of a person's life, statistically speaking. This, by the way, is another good argument against the idea that all aggression is learned. Most toddlers occasionally engage in reactive aggression, including punching, scratching, even biting, regardless of whether they have ever seen these behaviors modeled and without ever having been rewarded for them. Aggression is such a deeply rooted, primitive behavior that it doesn't need to be learned. So, thanks to their violent propensities, young children eventually learn what happens when they hurt someone: that someone gets upset. When people get hurt, they cry or cringe or act otherwise distressed. And just as is true for wolves, these sorts of behaviors are fairly effective at terminating aggression in normal children. In one study conducted in the 1970s, researchers examined the behavior of young children forced to share a box full of gerbils—the elementary school equivalent of heroin—with another child. The researchers set the box down between each pair of children, said, "Go!" then hustled from the room. They were right to hustle, as *441* separate conflicts over the gerbils erupted during the study, which involved only 72 pairs of children.

The outcome of each conflict was recorded by the researchers, who discovered that one of the best ways for a child in possession of the gerbils to keep another child from grabbing them away was to adopt what are called *oblique brows*—the raised and upwardly angled brows that are a key component of both fearful and sad expressions. They look like this: / \. Oblique brows were a more effective gerbil retention strategy than any attempts to use logic ("My turn!" "I have to feed him!") or physical force. Just as is true for wolves, the best way to resist attacks by gerbil-crazed six-year-olds is to use appeasement—to make them not want to attack.

The VIM has deterrent effects as well. As they develop and gain experience in social conflicts, children become able to predict in advance what sorts of behaviors will cause others to exhibit distress, and eventually they refrain from engaging in these behaviors with their peers. This is an essential part of the process of turning uncivilized toddlers into trustworthy members of their social group. The mechanism continues working throughout life. A recent study found that even during negotiations between adults, up to 12 percent more value can be accrued by a negotiator who expresses sadness as compared to anger or no emotion.

At least this is how things are supposed to proceed, but unfortunately, in a small percentage of children, the VIM doesn't work. Approximately 7 percent of children, or about one in fifteen, will qualify for a diagnosis of *conduct disorder* at some point during childhood. Children receive this diagnosis when they persistently engage in behaviors that are violent or cruel or otherwise violate the rights of others. The occasional schoolyard fight or squabble over gerbils doesn't count as conduct disorder. Children with this disorder threaten, bully, steal, and vandalize. They may set fires or engage in forcible sex. They are genuinely problematic.

Here are the full criteria for a conduct disorder diagnosis, according to the most recent *Diagnostic and Statistical Manual of Mental Disorders* (also known as *DSM-5*), published in 2013 by the American Psychiatric Association. Children with conduct disorder must have exhibited at least three of these fifteen criteria in the past year, with at least one criterion present in the past six months:

Aggression to People and Animals

1. Often bullies, threatens, or intimidates others.
2. Often initiates physical fights.
3. Has used a weapon that can cause serious physical harm to others (e.g., a bat, brick, broken bottle, knife, gun).
4. Has been physically cruel to people.
5. Has been physically cruel to animals.
6. Has stolen while confronting a victim (e.g., mugging, purse snatching, extortion, armed robbery).
7. Has forced someone into sexual activity.

Destruction of Property

8. Has deliberately engaged in fire setting with the intention of causing serious damage.
9. Has deliberately destroyed others' property (other than by fire setting).

Deceitfulness or Theft

10. Has broken into someone else's house, building, or car.
11. Often lies to obtain goods or favors or to avoid obligations (i.e., "cons" others).
12. Has stolen items of nontrivial value without confronting a victim (e.g., shoplifting, but without breaking and entering; forgery).

Serious Violations of Rules

13. Often stays out at night despite parental prohibitions, beginning before age 13 years.
14. Has run away from home overnight at least twice while living in the parental or parental surrogate home, or once without returning for a lengthy period.
15. Is often truant from school, beginning before age 13 years.

Obviously, any child who exhibits three or more of these behaviors is seriously troubled. But not all children with conduct disorder are troubled in the same way, or for the same reasons. Somewhere between half and two-thirds of children with conduct disorder exhibit primarily the reactive form of aggression. They are usually not deliberately cruel; rather, their fights and threats and destructiveness seem to be driven by fear or frustration. Importantly, they are also emotionally reactive *after* they act out. If they hurt someone or lose control, they may cry, wonder aloud what is wrong with them, or express unprompted remorse for what they have done. They seem genuinely sorry that their behavior may have hurt siblings or parents or friends they care about. These are the children for whom conduct disorder has most likely resulted from experiences of trauma or abuse, or perhaps an innately reactive or dysregulated temperament crossed with a moderately stressful environment. In these children, the Violence Inhibition Mechanism itself is probably intact, but its effects are sometimes overwhelmed by stronger forces. Focusing on these other forces, by either ameliorating sources of stress or trauma in the child's environment or treating symptoms of depression, anxiety, or lack of impulse control with medication or psychotherapy often causes conduct disorder symptoms to abate as well. This outcome suggests that these children's conduct disorder is a *secondary diagnosis* that is itself being caused by something else.

What about the other one-third to one-half of children with conduct disorder? For these 2 or 3 percent of all children, conduct disorder is not secondary to depression or anxiety. These children are, if anything, emotionally underreactive. Their aggression often isn't accompanied by anger or upset—sometimes it seems to come out of nowhere, and to be purposefully cruel. Worse, it isn't followed by any display of appropriate emotions like guilt or remorse in response to others' distress. Being unruffled by signs of the distress that their cruel or violent behavior has caused others makes these children especially worrisome because it suggests that, for them, the Violence Inhibition Mechanism itself is impaired.

One reason these children may not respond appropriately to others' distress is that they have trouble recognizing others' distress. Specifically, as James discovered, these children show a deficit that is the mirror image of what I had found in my own work: the least empathic children are also the worst at recognizing facial expressions of fear. When shown images of frightened faces like the ones I used in my graduate work, or when played recordings of frightened voices, they fail to recognize the faces and voices as fearful, and they fail to show the same empathic responses to them that healthy children show, such as an increase in sweat on the palms of the hands.

And these are the children who are most at risk for becoming psychopaths.

⁂

I vividly remember the day I met my first such child. In 2005, another research group at the NIH called to tell us that a boy enrolled in their protocol might be a better fit for ours. The other group initially thought that simple mood dysregulation might be his main problem because he had frequent temper tantrums. But mood dysregulation alone doesn't produce tantrums like Dylan's.*

First, Dylan was twelve at the time, an age when, for most children, actual tantrums are in the distant past. Normally the preschool years are the ones that brew the most tantrums. And while a two-year-old's tantrums may be annoying or frustrating, they are rarely significant problems. Now imagine an equally deranged tantrum produced by a boy who was five-foot-three and 120 pounds and able to reach every potential weapon in the house—knives, matches, baseball bats—and imagine it lasting an hour or more. Terrifying, right? That was Dylan. His bouts of rage would usually start over little nothings—irritation over not getting something he wanted or being punished for some misbehavior—and then quickly escalate, sometimes to red-faced screaming, other times to threatened or actual violence and destruction. During some of his worst episodes, he had threatened his parents with violence, punched or kicked holes in walls and doors, and, in one instance, smeared the walls of a room he'd been locked in with his own excrement. Once he threatened his mother with a knife. On that occasion and others, his mother brought his younger sisters to

---

*To protect the anonymity of individual participants, case study details have been combined to create composites, and names and identifying details have been changed.

a relative's house to spend the night for fear Dylan might actually follow through on his threats.

The details of Dylan's tantrums provided our first clue that something beyond simple mood dysregulation was afoot. When a person has a tantrum, it sometimes appears that he has completely lost control and might literally be capable of anything, so helpless is he to counter the forces of the emotional maelstrom. But this is an illusion. Tantrums can actually be experimentally induced in laboratory animals like cats and monkeys by stimulating a region of the brain called the medial hypothalamus. The old-school technique was to insert a tiny electrode into this small, evolutionarily ancient structure and turn on the current. When electricity is sent surging through the medial hypothalamus of a cat, it can send the cat into a snarling, spitting, clawing rage that looks for all the world like a toddler having a tantrum—but *only* if there is another living creature in the cat's environment. Rage, even electrically generated rage, needs a target. The same phenomenon has been demonstrated more recently using a much more precise technique, called optogenetics, in which the DNA of neurons is manipulated to make them fire in response to pulses of light. Then a tiny light-emitting optical fiber is embedded within the brain. When it is turned on, it will cause nearby genetically altered neurons to fire. Optogenetic triggering of neurons within the medial hypothalamus of a mouse will similarly cause it to launch coordinated, rageful attacks against other mice, or even against a wiggly rubber glove. But if the cage is empty when the light pulses—no aggression.

So even though a tree that falls in an empty forest will still make a noise, a child who trips and falls in an empty room will probably not throw a tantrum. What's the point? Rage is an emotion designed to make things happen, usually by cowing someone else into submission (as the angry wolf did). This is probably why, among animals like monkeys that observe clear social hierarchies, electrically induced rage will not be directed at just any living target, but mainly at targets who are lower-ranking than the raging monkey. Higher-ranking monkeys, who are likely to be unimpressed by a subordinate's rage attack, are usually spared. What does this mean? It means that even when an external machine is generating the rage attack—which would seem to be as uncontrollable as rage can get—the resulting behavior can still be modulated. The organism remains capable of maintaining some level of conscious or unconscious control over its behavior in accordance with basic biological rules—attack only living things, but not if they outrank you. This means that a child whose *only* problem is mood regulation will be very unlikely to go to extremes like threatening to stab his parents with a knife or smearing his own excrement on the walls, even in the grip of the most towering rage.

So what was going on with Dylan? The first step in finding out was to interview him.

The day of the interview I didn't know what to expect. It was my first clinical interview of any kind, much less one with a child with rage and violence issues like this one. Images from *The Silence of the Lambs* and *One Flew over the Cuckoo's Nest* flashed through my mind as I made my way through the daffodils outside Building 15K alongside my fellow postdoc and research partner, Liz Finger. Liz is a brilliant Harvard-trained neurologist who is insightful and perceptive, but she had as little prior experience with children like Dylan as I did. Together we walked toward the NIH Clinical Center, where Dylan was being housed in a locked ward, with mixed feelings of excitement and trepidation. Would Dylan be hostile? Would we be safe? Was he going to be restrained somehow? We were both smallish women, and prior to beginning the project we had received a rundown of basic safety procedures with potentially violent research subjects. Never allow any loose pens, pencils, or other potential weapons near the subject. Stay at least three feet away, out of arm's reach. Never let the patient get between yourself and the door. Don't turn your back. We hoped these feeble measures wouldn't be our only source of protection.

We made our way through the Clinical Center's gleaming metal-and-glass atrium, weaving through the physicians and patients streaming through it, the physicians' brisk and purposeful gaits in stark contrast to the patients, who clung unsteadily to their IV racks or slumped in chrome-plated NIH-issue wheelchairs. We hung a right and came to the pediatric inpatient ward, where, after a moment's wait, we were buzzed in. "Here to see Dylan?" said the friendly, round-cheeked nurse who greeted us at the door. "Follow me."

Toward the end of the hallway she pointed at a nondescript, blond wood door, one in a long succession of identical doors, and said, smiling, "Here you go!" before departing. We hesitated a moment, then turned the handle and walked in.

The room was small and spare, but cheerful and tastefully decorated in the desert tones of a modern hospital. Dylan perched on the corner of a neatly made twin bed in a posture that suggested he had been waiting for us. I hoped I didn't look as taken aback as I felt when I saw him. I briefly thought the nurse had gotten the room wrong. *This* was the Dylan whose files we'd been poring over? This was the boy whose parents and sisters lived in fear of him? The knife-wielding menace? The shit-smearer? This boy looked like he'd wandered in off the set of a cereal commercial. He was gangly-legged and suntanned, with a shock of white-blond hair and a nose spattered with freckles. He stood up politely to shake our hands, evidently well practiced in the art of greeting strange adults. When he smiled, it was such a broad, open smile, so incongruous with everything we had heard and read about him, that I couldn't help myself. I liked him at once.

I never stopped liking him. We had a perfectly lovely conversation that day, as we did every time we met afterward. Dylan told us about his home in Arizona, about his love of playing golf with his mother, whom we met that day as well. She was also a dazzling mosaic of tan skin and gleaming teeth and expensive golf clothes, and her affection for Dylan was clear. In the snug confines of that hospital room, with its soft colors and shining maple laminate floor, keeping myself between Dylan and the door was the last thing on my mind. My abiding impression was one of friendliness and warmth.

Liz and I spoke to Dylan in private for a while, and he confirmed that he had indeed done all the things we had heard about. But the way he explained everything to us, all the wild behaviors somehow lost their sharp edges. Every explosion was just the inevitable outcome of a bad day—he was tired, he was frustrated, his sisters had been bothering him. He never really meant to hurt anybody. He didn't know why he threatened people, or why they believed he was serious when he did. He just got upset. All his explanations seemed so reassuringly normal. If anything seemed out of the ordinary, it was that Dylan seemed a little more fidgety than the average adolescent boy, changing positions often, animated by a restless energy that comported with the nurses' observations of his impulsiveness and difficulty concentrating, but that was the extent of it.

We left the interview shaking our heads. Whatever we had been expecting to find that day, we hadn't found it. He seemed like *such* a nice kid. And the nurses we spoke to after we interviewed Dylan and his mother (separately) said that he was often sweet to the younger children on the ward—reading to them or helping them with their schoolwork.

Dylan had just demonstrated to us why using an interview alone to evaluate psychopathy is a bad idea.

The modern clinical definition of psychopathy is largely based on the work of the twentieth-century psychiatrist Hervey Cleckley, which he detailed in his masterful book *The Mask of Sanity*. The text is a wide-ranging exploration of the meaning of sanity and insanity, morality and immorality, and includes fifteen sharply observed case studies that illustrate how psychopathy is distinct from other psychiatric disorders. After presenting his case studies, Cleckley summarizes the essential characteristics of psychopathy. He begins with an observation that echoes Tony Savage's description of Gary Ridgway, the Green River Killer:

> More often than not, the typical psychopath will seem particularly agreeable and make a distinctly positive impression when he is first encountered. Alert and friendly in his attitude, he is easy to talk with and seems to have a good many genuine interests. There is nothing at all odd or queer about him, and in every respect he tends to embody the concept of a well-adjusted, happy person. Nor does he, on the other hand, seem to be artificially exerting himself like one who

> is covering up or who wants to sell you a bill of goods. He would seldom be confused with the professional backslapper or someone who is trying to ingratiate himself for a concealed purpose. Signs of affectation or excessive affability are not characteristic. He looks like the real thing.

Cleckley could not have described Dylan better if he had been sitting in on the interview with Liz and me. Not simply *less* maladjusted than the average child in a locked psychiatric inpatient ward, Dylan genuinely came across as a friendly, normal, well-adjusted twelve-year-old. This stark contrast between his frequent threats and violence and his outwardly friendly and normal appearance was the second clue that Dylan's problem was not simply poorly regulated moods. Together, these two pieces of information—unusually violent behavior even for a psychiatric patient, as evidenced by his files, and a hypernormal, even charming, outward appearance that betrayed no hint of how violent he could be—suggested that Dylan might be psychopathic.

The concept of a psychopathic child makes many people queasy. In some ways, the two categories seem mutually incompatible. Children, even badly behaved ones, are viewed as maintaining some fundamental innocence compared to adults, whereas psychopaths are viewed as fundamentally depraved. But of course, neither stereotype is totally true. Children, just like adults, are capable of cruelty and violence, and even highly psychopathic people are not cruel or violent all of the time. Perhaps our resistance to the idea of a person being both a child and psychopathic—of there being overlap between these two groups—reflects our moral typecasting biases, according to which children fill the role of moral patients and psychopaths fill the role of moral agents.

But in reality, psychopathy is a developmental disorder. It does not emerge out of nowhere in adulthood. Essentially, without exception, all psychopathic adults first showed signs of psychopathy during adolescence or childhood. This means that for every psychopathic adult out there in the world, there was once a psychopathic child.

The title of a widely circulated *New York Times Magazine* article from 2012 posed a highly provocative question: "Can you call a 9-year-old a psychopath?" The question is not so provocative from a scientific perspective. The definition of an adult psychopath is anyone who meets a specific cutoff on a measure called the Psychopathy Checklist—Revised (PCL-R), a 40-point scale scored using information from an interview and background files. In the United States, an adult who scores at least 30 points out of 40 is deemed a psychopath. This is a debatable practice, as there is no functional difference between people who score 31 versus those who score 29 (not to mention that any two assessments of the same person can and usually do differ by at least two points), but it remains the current standard. There is a nearly identical 40-point scale designed for use in children as young as ten called the Psychopathy Checklist: Youth Version (PCL:YV). It is possible for a nine-year-old to possess all the personality and behavioral traits that would lead us to label an adult a psychopath, and such nine-year-olds often do go on to become adult psychopaths.

But from a broader cultural and moral perspective, the answer to the question "can you call a nine-year-old a psychopath?" is: absolutely not. The terrible stigma that results from labeling a child a psychopath cannot be ignored. And although every adult psychopath began as a psychopathic child, the reverse is not true: many children with high psychopathy scores do not go on to become adult psychopaths. Why is not entirely clear. We know very little about how the brain develops during adolescence, and some children may genuinely remit as their brains rewire themselves in the period leading up to adulthood. Remission could occur in response to favorable changes in a child's environment, or perhaps as a result of innate developmental processes. Other children who appear to remit were probably misclassified to begin with, such that what looked like emerging psychopathy was instead an unusual expression of early bipolar disorder, schizophrenia, or even autism. For these reasons, no responsible researcher or clinician will ever label a child a "psychopath." Making this rule easy to follow is the fact that there is no official cutoff score on the PCL:YV, which removes the temptation to pin a label on a child who might well turn out to be something quite different. You will never hear me call any of the children we worked with a psychopath. They simply were not.

But the fact that children can strongly express psychopathic *traits* cannot and should not be ignored. So researchers and clinicians try to split the difference by referring to such children as possessing psychopathic traits or tendencies—or, for concision, as just "psychopathic" ("psychopath-ish" not being a real word). Often the phrase *callous-unemotional traits* is also used as an even less incendiary description of the key personality traits that typify psychopathy. *DSM-5* eschews mention of psychopathy entirely, but there is a new designation reserved for children who possess the antisocial and callous features of psychopathy, which is the inelegant but accurate designation *conduct disorder with limited prosocial emotions*. This designation is satisfied if a child has conduct disorder and also exhibits at least two of four key characteristics across different settings: lack of remorse or guilt, callousness (a lack of empathic concern), lack of concern about performance in important activities like school or work, and shallow or deficient affect. This specifier did not yet exist when we were beginning our research. So, for our purposes, any child with a diagnosis of either conduct disorder or its developmental precursor, *oppositional defiant disorder,* and a PCL:YV score of at least 20 was deemed to possess sufficient psychopathic traits to qualify for our study.

Was Dylan such a child? Liz and I each evaluated him separately using the PCL:YV, which we had been trained to administer by David Kosson and Adelle Forth, two of the psychopathy experts who created the scale. We took into account both his behavior during the interview and all of the background information we had collected about him. Dylan scored a 0 on a few items on the scale, including "serious violations of probation," since he had never been on probation, and "grandiose sense of self-worth," since he seemed merely confident, not narcissistic or grandiose. But as we went through the scale, his score continued to mount. "Early behavior problems?" Yes.

"Poor anger control?" Definitely yes. "Impression management?" Also yes—among other things, during his interview he consistently angled to portray himself in the best possible light, despite our already knowing the facts of his background. "Failure to accept responsibility?" Interesting—also yes. Despite the favorable impression we had gotten of Dylan overall, when we reviewed our interview notes it was clear from both Dylan's and his mother's accounts that Dylan never accepted any responsibility for his behavior. Everything was always the fault of some external factor—a bad day, someone else bothering him. The same was true for "lacks remorse." He'd had ample opportunities to express remorse for the effects of his actions on others, but he never really did. Instead, he minimized the seriousness of his actions and blamed them on others rather than acknowledging how much distress he had caused his family and teachers.

Liz tallied up our scores when we were done, reaching almost perfect agreement in our assessment of Dylan. He scored a 24. We had our first research subject.

⁂

That marked the beginning of a long, often grueling process of recruiting several dozen children with psychopathic traits for our brain imaging studies. A few, like Dylan, were sent our way by other investigators at the NIH, but most we had to find ourselves. If you assume that at least one child in 100 would score at least 20 on the PCL:YV (which is, if anything, a lowball estimate), a metropolitan area the size of Washington, DC, contains thousands of potentially eligible children. But recruiting them is a difficult business. There are no advocacy organizations or listservs for parents of children with psychopathic traits like there are for parents of children with autism or ADHD. So we had to create advertisements. But they had to be delicately worded. "Is your child psychopathic?" was not going to fly. Not only would it be inflammatory, but many parents of eligible children don't think of their children in these terms (although some do). So we instead asked parents about their children's behavior. Our advertisements, which we posted in print media and as posters near family courts and probation facilities, asked, "Does your child have behavior problems and not feel guilty when he/she does something wrong?" These ads yielded very few calls relative to the number of potentially eligible kids out there, perhaps because most parents with a child who met these criteria were already stretched thin—perhaps too thin to participate in anything that didn't hold out the promise of treatment. But eventually, calls from parents began to trickle in.

If you have ever felt compassion for anyone in your life, feel compassion for the parents of these children. During their initial phone screenings, and later during interviews in the lab, the stories these parents told us about their children were heart-rending. By the time they called us, their children's misbehavior had usually been going on for many years. There was rarely a single calm day in their homes. Like Dylan's parents, they worried every day about what new episode of violence or theft or destructiveness the day would bring, about the safety of their other children, and about their own safety as well.

Several of them had been seriously injured by their children. One mother told us about her son shoving her so hard during a fight that she broke her wrist when she fell. A father told us that his preteen daughter had kicked him in the face with such force that he feared he might lose his vision in one eye. What had the father done to earn such a kick? He had no idea. He'd been sitting on the floor watching television at the time. We heard from parents whose children stole from them constantly—collectively hundreds or thousands of dollars. No matter where cash and valuables and credit cards were hidden, it never seemed to stop the thievery. That wasn't even counting all the costs the parents incurred from damage to their possessions—wrecked cars, dead pets, fires. The parents were lied to. Manipulated. Subject to endless haranguing by school teachers and administrators fed up with the children's misbehavior in school—a problem that the parents were even more powerless to fix than the misbehavior at home. Most of the children had been thrown out of at least one school, sometimes several, often for injuring classmates or teachers. One mother told us that her daughter brought a glass bottle of juice to school specifically so that, when she finished the drink, she could smash the empty bottle across the face of a teacher she disliked, which she did, right in the front of the classroom, leaving a gash that required seven stitches to repair. The mother of another child had had to retrieve her son from school so often when he was suspended—at least a dozen times—that she ended up being fired from her job for lost work. She was a single mother, and the strain of it all was so severe that she was briefly institutionalized.

I remember that interview mainly owing to her son Michael's response when I probed a bit to ask how he felt about all the trouble he had caused his mother. Did he feel badly that he had caused her so much suffering? I was curious in part because, much like Dylan, Michael seemed to have a very sweet relationship with his mother when they visited us. The question stumped him, though. I think he realized he should feel some sense of remorse or guilt, but somehow he couldn't conjure up the feeling I was asking about. Finally, he said, "Well, the things I do hurt her, right? But she doesn't say how much, so it doesn't really have an effect on me." Michael and Dylan had little in common other than both being adolescent boys. But one clear commonality was that no matter how much hurt and distress they caused other people, it didn't seem to occur to them to feel badly about it.

As we met more and more of these children and their parents, our ability to evaluate them continued to improve, although some cases were easier than others. Perhaps our most clear-cut case was Jamie, a sandy-haired, button-nosed twelve-year-old boy who bounced in, visibly crackling with energy and trailed by his kindly but beleaguered-looking mother. We always started out our interview sessions by talking to the parent alone. We would interview the child alone next, a sequence that allowed us to spot when a child was lying or glossing over misbehaviors. With Jamie, there was little need for this sequence. He couldn't have been prouder to detail his many, many exploits to us. His misbehaviors ran the gamut. Other than the items related to serious crimes and sexual offenses, he received top marks across the PCL:YV. He stole things. He set fires. He lied. He charmed and manipulated. He constructed elaborate cons to bilk his schoolmates of their money or possessions. Despite only being in middle school, he was running a successful loan shark operation out of his bedroom in a wealthy suburb of Richmond, Virginia. Interest payments ran to a dollar a day. When payments ran late, Jamie threatened to shoot fireworks at his clients. Many of these clients were high school boys who must have towered over him, but they took him seriously. Among Jamie's more florid exploits was the time he somehow procured an artificial grenade and lobbed it into his local public library, "just to see how people would react." Not surprisingly, people reacted by fleeing the building. Crying, screaming children and their parents poured out in a panic. While everyone else ran away from the building, Jamie and his friend ran *toward* it—carrying a video camera, no less, the better to record the terror and mayhem they had caused for posterity. Jamie's pride in his caper was obvious. "It was," he concluded with a crooked grin, "a total Kodak moment."

Jamie wasn't our highest-scoring subject, though. That would be fourteen-year-old Amber. Amber panthered into our interview room oozing charisma and sexuality that made me, a thirty-year-old woman at the time, feel unsettled. I can only imagine what her effect on young boys and men was. She was fully aware of this. Like many of the girls we worked with, Amber had figured out that it was often easier to get what she wanted using charm and allure than threats or violence. Like anyone else, children with psychopathic traits use the tools available to them to get what they want. They just care less than others about the collateral damage those tools may inflict. Amber liked expensive clothes, for example, and older boys and men were the ones with the means to buy them, so she seduced them. She didn't much mind that they would end up branded as sex offenders for life if they slept with her.

Amber had the highest intelligence score of any of our study participants, and it showed. In typical kids, a higher IQ tends to be associated with fewer behavior problems, but in children with psychopathic traits the opposite seems to be true—a higher IQ coupled with a psychopathic personality seems to lead to more serious offenses, perhaps because the combination yields a sort of canniness that helps them get away with ever-ascending misdeeds without getting caught. Amber was certainly preternaturally perceptive for a young teenager. I could feel myself being sized up throughout our interview—she watched our faces keenly as she fed us stories about killing the family's pet guinea pig or threatening to burn down her family's home while they slept. Like Jamie, she wasn't trying to hide anything. Quite the opposite. She was happy to describe what was going on in her head during these exploits. She explained, for example, how she sized up adults to avoid being punished. "Some adults are impressed when I use big words," she said. "Others will let me off if I cry." Her mother confirmed that Amber could produce effective crocodile tears. Once, her mother had found in Amber's bag a printout from the Internet titled "A Guide to Shoplifting," along with a pile of cosmetics and an expensive handbag from an upscale department store. When confronted, Amber had burst into tears, protesting that she was sorry and she would never do it again. Her mother admitted that she'd been mollified by similar displays in the past, but was so galled by the brazenness of this incident that she snapped, "Oh, get real." And Amber's tears, she said, stopped like a switch had been flipped, replaced by Amber's usual level stare. Amber was the only one of the children we tested with whom I would have been unwilling to spend a night alone in a house.

Not every child we talked to was so frank about their motivations and behaviors. Heather was among our most difficult subjects to evaluate. According to her father, she was a terror. Like Dylan, she threw wild, violent tantrums that sometimes went on for hours and left her limp and spent. She also engaged in proactive forms of aggression that were just subtle enough that she could claim she hadn't been doing anything wrong. Her father suffered frequent migraines, and Heather went out of her way to slam doors and flip on bright lights whenever one of his headaches came on, seemingly getting a visible kick out of the pain these sensory assaults caused. She was sometimes violent at school, once striking a schoolmate with a toy with such force that he required stitches. And like many of our other study participants, she manipulated, lied, stole things, and lied about stealing things.

At least, according to her father all this was true. But when Liz and I sat down to talk to Heather, we were stunned by how differently she came across in person. Heather had the limpid brown eyes and long limbs of a doe, a sweet, shy smile, and a soft voice in which she told us stories that would begin in the same place as her father's stories, but always ended up somewhere completely different—inevitably a place in which Heather had committed no wrongdoing. It was her father who had the terrible temper and was forever taking out his unhappiness on Heather (said Heather). And when the interview ended, I watched in amazement as Heather carefully cleaned up the wrappers and crumbs from the food she had been eating while we spoke. Liz and I emerged from the interview stumped. It was by far the biggest mismatch between a parent's and a child's stories we'd encountered. Even knowing how genuinely winning many of the children we met could be, it was hard to understand how Heather fit into our study. We ended up calling a referee—in this case, one of Heather's teachers. We asked for details on several of the stories we'd heard two versions of that the teacher had personally witnessed, and in every case the teacher's stories echoed those of Heather's father. Heather was just an incredibly skilled deceiver—the best we ever encountered. If we had used just her interview, we would never have pegged her for one of the children we were studying. I could easily imagine having offered her a research assistantship or a babysitting position based on her interview. But when we tallied her up, she scored well above our cutoff. Yet another lesson learned.

⁂

I can guess what you're thinking at this point. I have talked to many people about these children and their families over the years and tend to hear the same comments over and over. In some recess of your mind, the thought *These kids must have* really terrible *parents* may be bouncing around. The belief that badly behaved children are the product of bad parenting is so deeply rooted in our culture that it is difficult to dispel. But let me try. First of all, we acquired a *lot* of information about these families during the course of our screening and interviews, and while they varied in many ways, a common thread was that the parents had tried literally every possible option to help their children before coming to see us—counselors, medication, special schools, social workers. These were caring parents with resources who really were trying, but none of it helped. I'm not saying they all were the best parents in the world (they varied, naturally), but they were definitely not so bad that their parenting alone could have produced such children.

As evidence, nearly all of these parents had other children, most of whom we met, and none of whom were also psychopathic—much like Gary Ridgway's normal siblings. If unusually terribly behaved children are the product of unusually terrible parents, then the children of such parents should uniformly be a mess, right? But they're not. It's not that poor parenting can't result in badly behaved children—of course it can. But it doesn't make children psychopathic. Parents who are overly permissive or simply unskilled may end up with ill-mannered or entitled or bratty children, but these problems can often be solved with a little coaching on setting clear limits and not rewarding misbehavior. And parents in highly dysfunctional households marked by domestic violence or neglect or abuse may produce children with significant behavior problems. But again, most of these children's aggression tends to be of the emotional, reactive variety, and they often present with depression or anxiety as well.

When the misbehaviors are of the purposefully cruel, manipulative, deceitful, or remorseless variety, it's a different story. Engaging in these kinds of behaviors seems to be much more strongly driven by inherited factors, as we know from adoption and twin studies. Recall that these studies consistently show that parenting and other environmental factors explain only a small fraction of the variation in the proactive aggression that psychopathic children engage in.

I should note that some recent studies examining the relationship between child psychopathy and parenting find that children with higher levels of psychopathic traits have parents who are colder or more neglectful than average. The urge to assume that the arrow of causation runs in a straight line from cold and neglectful parents to callous and remorseless children is a powerful one. But there are several alternatives. One is that cold and uncaring parents end up with cold and uncaring children because they share similar genes that predispose them toward this personality profile. More than one adoption study suggests this is true. Another alternative is that a child who is difficult from the get-go causes his parents to become colder, less attentive, quicker to punish, and harsher when they do, a dynamic also backed up by research. These various causal pathways may of course interact as well, such that different styles of parenting may buffer or exacerbate the expression of psychopathic traits. For example, some recent research suggests that very high levels of parental warmth may lessen the severity of inherited psychopathic traits. So although cold parenting does not *cause* children to become psychopathic, interventions focusing on increasing the warmth of parents' interactions with their children may help to reduce symptoms, particularly in younger children.

The parents of the psychopathic children we studied, however, had never had any of these nuances explained to them. Instead, everyone from pediatricians to school principals to neighbors had placed the blame squarely on them and their putatively rotten parenting. They often blamed themselves as well, being subject to the same cultural forces as the rest of us. More than one parent became teary when we asked if they thought their son or daughter ever felt remorse for any of the things they'd done. When I asked Michael's mother this question, her face crumpled. After a long silence, she responded, "I want to think he does, but . . . ," then trailed off.

What could I say? She was right. He didn't care. Many days my heart ached for these children's parents long after I left the interview room.

⁎

So what *was* going wrong with these children? Our goal was to find out using the (at the time) fairly new technology called functional magnetic resonance imaging (or fMRI for short) to peek inside the active brains of these children. The emergence of fMRI in the 1990s revolutionized the field of cognitive neuroscience, which aims to identify biological mechanisms that undergird mental processes like attention, memory, and emotion. Before the emergence of fMRI, researchers who wanted to identify a malfunctioning brain area in a clinical population had only a few choices. One was to use positron emission tomography (PET), which requires injecting study volunteers with a radioactive sugar solution, then hustling them into a PET scanner before the radioactive isotopes decay. The end result is fairly fuzzy images of energy consumption levels inside the brain. Or, if researchers suspected that the problem lay in some dysfunction in the brain's cortex, right underneath the skull, they could use electroencephalography (EEG) to measure electrical potentials across the scalp. But, like PET, the readings produced by EEG are spatially fuzzy, making it hard to tell exactly what part of the brain is generating the signal.

Although both PET and EEG are valuable, fMRI opened up new worlds to cognitive neuroscientists, permitting direct and spatially precise measurement of activity deep inside the brains of living, behaving humans. An MRI scanner is just a giant, doughnut-shaped magnet; fMRI is the use of such a magnet to detect small increases in the flow of blood to brain areas that are active and clamoring hungrily for the fuel that blood carries. Unlike a PET scan, an MRI scan uses no radiation, although it does have other limitations, which mainly stem from the fact that fMRI measures blood flow rather than activity in neurons themselves. Measuring blood flow limits the temporal precision of fMRI because the ebbing and flowing of blood in the brain (termed the *hemodynamic response*) takes a few seconds, whereas changes in actual neural activity take place in milliseconds. On the other hand, fMRI's spatial precision is good and getting continually better as stronger magnets and more advanced software are rolled out. The MRI machine we initially used at NIMH in 2004 was a 1.5-Tesla magnet, which is about 50 percent stronger than the magnets that lift cars in scrapyards. Later we switched to a 3-Tesla magnet, which is standard today. Recently, the NIMH acquired a 7-Tesla magnet, which is so powerful that it can measure changes in brain activity with a spatial resolution of one cubic millimeter—although it wreaks such havoc on charged particles in the inner ear that volunteers must be rolled into its bore very slowly to avoid getting vertigo or vomiting. I had the occasional bout of dizziness working next to the 3-Tesla magnets, but the sensation wasn't unpleasant, just strange (and not nearly as distracting as its sly tugs on the metal hooks and rings of my bra).

We would be using fMRI to measure activity in parts of the brains of psychopathic teenagers that are hard to assess using any other approach, in particular the underside of the prefrontal cortex, which sits right above the eyes, and a region called the amygdala. The amygdala (Latin for "almond") is a lump of fat and fiber about half an inch in diameter that is buried beneath layers of cortex under each temple. The structure is so small and lies so far beneath the scalp that neither PET nor EEG can reliably measure its activity. But its small size belies its importance. Among other things, it plays a critical role in recognizing fearful facial expressions.

This had first been discovered in 1994, prior to the advent of fMRI, through neurological studies of a patient with a very rare kind of brain damage: total obliteration of both the left and the right amygdala, and nothing else. No accident or stroke can do such precisely localized damage; instead, the culprit is a rare genetic illness called Urbach-Wiethe, which can cause the amygdala to gradually calcify over the first decade or two of a person's life. In the late 1980s, a group of researchers, led by Daniel Tranel at the University of Iowa, was approached by a woman with this condition, whom they called S.M. to protect her privacy. S.M. was twenty at the time, with a pleasant, open face, a breathy voice, and a flirtatious, disinhibited demeanor. She liked to stand about twelve inches away from the person she was talking to, and the researchers' first published description of her drily described her "tendency to become somewhat coquettish" during her testing sessions. A CAT scan of her brain confirmed that her amygdalae were totally destroyed. Intrigued, the researchers ran S.M. through dozens of cognitive tests to see what else she'd lost along with this structure.

Many of her mental abilities remained intact, including her intelligence and memory, but among the deficits the researchers uncovered was S.M.'s inability to recognize others' fear. The researchers presented her with a series of emotional facial expressions, including some of the same ones I'd used in my research, and asked her to provide a label for each. She had no trouble with faces that expressed anger or disgust or happiness or sadness; her performance corresponded very well to that of other adults, including adults with damage to other parts of their brains. But when she was shown photographs of people who looked frightened, she bottomed out, describing the expressions alternately as sad, disgusted, angry—nearly anything *but* fearful.

The researchers wondered what S.M. thought fear *did* look like, so they asked her to try to draw a frightened face, along with faces expressing other emotions, a request that revealed her knack for portraiture. Her depictions of anger, sadness, and disgust were vivid and easily recognizable. The angry face glowered fiercely, looking a bit like a bearded Fidel Castro in his prime. Tears dripped from each eye of the sad face, the eyebrows of which were perfectly oblique. But when asked to draw a fearful face, S.M. literally drew a blank. She protested that she had no idea what it would look like—no image at all came to her mind. She tried several times to create drawings that she ended up scrapping. Finally, she produced an image that looked nothing like the others. It was not a portrait, but a small, round figure shown in profile on its hands and knees. Its expression was hard to read, but it did not look frightened. Its mouth was closed, and its brows sat low and straight over its eyes.

Subsequent tests of other individuals with localized damage to the amygdala have consistently revealed similar patterns—their ability to recognize fearful facial expressions is reliably impaired. The most recent such study found fear recognition deficits nearly identical to S.M.'s in a teenage Urbach-Wiethe patient in Iran. Such patients sometimes have trouble recognizing fear from other cues as well, including vocal utterances, body postures, even emotionally stirring music. Damage to no other brain structure results in this specific pattern. These data make clear that the amygdala must play some important role in our ability to recognize expressions of fear.

Strikingly, they also point to the conclusion that psychopathic children's struggle to recognize others' fear may similarly stem from dysfunction in their amygdala.

To determine whether Dylan and Amber and Michael and the other adolescents we were recruiting in fact suffered from amygdala dysfunction, we needed to measure activity in this structure while they viewed fearful facial expressions. Viewing fearful expressions generally produces a robust amygdala response in healthy people. Hundreds of studies have now been conducted examining healthy adults' brain responses to facial expressions using fMRI, and this is what they reliably show. The amygdala is more active when people view fearful expressions than when they view any other expression—backing up what studies of S.M. and other Urbach-Wiethe patients suggest, which is that the amygdala plays some special role in processing this expression.

So every time we interviewed a newly recruited adolescent for our study, Liz and I would race back to score their PCL:YV and sign up those who were eligible for a brain scan as soon as possible. Speed was imperative. We were always in a race with the unknowable, unpredictable disasters that tail kids with psychopathic traits. More than one child who cleared all our screening hurdles—getting a conduct disorder or oppositional defiant disorder diagnosis, demonstrating a normal-range IQ, receiving high psychopathy scores—became unscannable soon afterward. Several ended up hospitalized or in detention. A few of the girls got pregnant. Occasionally parents threw in the towel and sent their children to live with another relative who might be better able to manage them. We thought we were in the clear on our first attempt to scan a boy named Derek, as his scan was scheduled only a week after his interview, only to have him lope in on scanning day wearing a bulky, metal-laced monitoring bracelet on his ankle that had definitely not been there before.

"What is that?" I asked in a panic.

"It's my monitoring bracelet," he said. "I just got it."

"Um . . . I don't think we can put you in the scanner wearing that, Derek."

"I can take it off."

"No, no, no, no. Please don't do that!" I hastened to say. "Why don't you just come back after it's off?"

Although MRI technology itself is quite safe, the presence of magnetic metal inside the scanning room can be catastrophic. Rarely, people have been injured or killed when loose metal oxygen tanks, scissors, or other objects were accidentally brought too near the magnet. These objects can be pulled into the bore with such force that they essentially become cannonballs in reverse, with dire consequences for anyone who happens to be inside. Even metal affixed to a person isn't necessarily safe. A firefighter once got sucked into an MRI via the metal oxygen tank strapped to his back. He ended up pinned inside the scanner with his knees pressed so hard against his chest that he was on the verge of suffocating by the time someone quenched the magnet. The metal checks we gave each child before a scan were the stuff of a TSA supervisor's dreams. Pockets were emptied, hair was checked and rechecked for pins and clips, jewelry was removed—even shoes, which sometimes contain a metal shank, had to come off. How quickly I came to loathe cargo pants, with their dozens of tiny pockets caching forgotten keys and safety pins.

We nearly had to cancel one scan when a subject named Brianna arrived wearing a new nose ring that she didn't know how to remove. Putting her in with the ring was not an option. It was a great, fat steel thing, and the scanner could have torn it right through her nostril. But as luck would have it, we were running two scans that day, right next door to each other, and the other participant was Amber, who had ample experience with piercings and who offered to help when she heard Liz and me conferring.

I remember looking at Liz in alarm. I knew what both girls were capable of. They were two of our most violent participants. I half expected that just being in the same room together would cause them both to explode, like matter and antimatter colliding. But we really wanted to scan Brianna before anything else came up. (As it happened, she became pregnant not long after.) And it wasn't like there was an *explicit* rule against letting participants take each other's piercings out. True to form, they were very polite with each other. I held my breath as I watched Amber unscrew the ball at the end of the nose ring and thread the ring carefully through Brianna's nostril. "Here you go!" she said, dropping the slightly damp ring into my palm. It was a sign of how grateful I was that it never occurred to me to be grossed out.

So, thanks to Amber, into the scanner went Brianna. Later, in went Amber. Dylan and Michael and Jamie and other teenagers like them followed. Inside the scanner, each watched through a mirror as a series of black-and-white fearful, angry, and neutral facial expressions flickered across a projection screen. I would love to know what was going through their minds as they watched. In many fMRI tasks, this is a critical question. If a participant spends the scan wondering what the task is about, the scan will be ruined—unless the researchers are trying to measure the brain activity associated with "wondering." We were running what is called a passive viewing task. Emotional facial expressions are such primitive stimuli that people don't need to consciously focus on them for their brains to respond. As long as their attention doesn't get too derailed, it's actually *better* if they don't focus on what the emotion is. The simple act of labeling an emotion, whether your own or someone else's, is actually a mild form of emotion regulation and can reduce the affective response to it. So we asked our participants to label the gender of each face instead. For over twenty minutes at a stretch, each child lay there pressing buttons to categorize over a hundred grimacing faces as male or female. It was such a deeply boring task that it's a wonder any of them finished it. Did I mention that they had to lie perfectly still the entire time? As little as four millimeters of movement—even just a jiggling foot—could render a scan unusable. Thank goodness we could pay them. Even a child with serious behavior problems can (usually) handle twenty minutes of stock-still boredom if $75 is waiting on the other side.

For years Liz and I spent every other Sunday under the glare and buzz of the fluorescent lights in the NIH basement collecting MRI scans of these children, as well as scans of matched healthy control children to whom we could compare them. All this work was of course in part aimed at reaching our own goals—our findings ultimately were published in prestigious psychiatry journals, which helped us later attain faculty positions, research funding, and invitations to conferences. But we were also strongly motivated by the broader gains our work could yield. Our studies were the first effort to directly measure brain dysfunctions that might contribute to psychopathic traits in children and adolescents and would represent an important step toward understanding the roots of psychopathy, which might lead to better ways of identifying and treating such children in the future.

Such steps are desperately needed. Children with conduct disorder and oppositional defiant disorder are every bit as mentally ill as children with bipolar disorder or anxiety or autism, but much less is known about the origins of their disorders because only a tiny fraction of available public and private research resources are devoted to understanding them—far, far less than the resources spent on relatively less prevalent and severe disorders. Given this, it's a small wonder that so few effective therapies—either pharmaceutical or behavioral—exist to treat them. This fact makes life worse for everyone: certainly for affected children and their families, but also for their friends, teachers, and other community members whose lives are negatively affected by the untreated aggressive and disruptive children in their midst. Despite how difficult and occasionally dispiriting it can be to study these children, the urgent need of these children and their families and communities remains a constant motivator.

At last we collected usable data from twelve non-pregnant, non-institutionalized, non-jiggly children with psychopathic traits and twenty-four matched controls (twelve of whom were healthy and the other twelve of whom had only ADHD). Subsequent analyses of the data took weeks because of the enormous amount of information that a whole-brain fMRI scan collects and the number of transformations the data must be subjected to before they can be analyzed. At the end of this long process, I finally conducted the statistical test comparing activation in the amygdala across the three groups of children while they viewed fearful expressions. When the analysis finished running, I opened up the image that would show me whether our hypotheses were confirmed: whether children with psychopathic traits fail to show appropriate elevations in amygdala activity when looking at these expressions. I scrolled through the image, the whorls of the cortex unfolding as my cursor moved deep into the temporal lobe, holding my breath until I got to the amygdala, hoping that it would show a cluster of differential activation—and there it was! A little glowing red blob signifying differences in activity across the groups of children, right where it should be.

On average, our psychopathic children showed no activation—zero—in the right amygdala when they viewed the face of someone experiencing intense fear as compared to a neutral face. The sight of another person in distress made no mark on this part of their brains. This was quite unlike what we saw in the healthy children and the children with ADHD, who, on average, showed reliable increases in amygdala activity, just as most adults do. Our finding, which has now been reproduced several times by researchers from different laboratories, helps explain why children with psychopathic traits have so much difficulty recognizing fear in other people—why the sight of the distress that their violence and threats cause others has so little power to inhibit their cruelty. It is because the region of the brain that is critical for accurately identifying and responding to these expressions is defective; as a result, these children literally struggle to understand what they are looking at.

More insight into this pattern of results emerged from the results of one of the cognitive tests that Liz and I had also been conducting with the children along the way. The test was aimed at evaluating the children's subjective experiences of different emotions. First we asked the children to recall times in their lives when they had experienced strong emotions themselves, including anger, disgust, fear, happiness, and sadness. Next, they were to describe details of each event and how it made them feel, in terms of both body sensations and psychological experiences. People who are psychopathic are known not to show strong physiological responses, like changes in sweating and heart rate, in response to images or sounds that most people find frightening. But no one had yet systematically inquired whether children with psychopathic traits *feel* fear psychologically in the same way as other children. As we discovered, they don't.

Overall, the children with psychopathic traits reported that they felt fear only infrequently and weakly. When asked, for example, how often they felt fear on a scale from 1 to 7, the average response for healthy children was a little over 4. Michael and Amber both circled "1" ("never"). Their responses echoed the stories we'd heard during our interviews. Michael was forever hurting himself when attempting physical stunts like riding his bicycle off the roof of his school; Amber's mother recounted in wonder that when Amber was in preschool she would sometimes run off and her mother would find her playing alone in the pitch-dark, spooky basement of their building. Some of the children with psychopathic traits reported that they had felt fear when, for example, they found themselves on a roller coaster that got stuck, or saw a falling tree narrowly miss their house during a hurricane. But when we queried them on how this fear felt, they reported not feeling the same intense physical changes as the healthy children, like muscle tension, shaking, or breathing changes. Two of the psychopathic children we queried claimed that they had *never* felt fear in their entire lives, whereas no healthy children said this.

This might be my favorite response to a question about fearfulness in the children I've worked with. This child, a thirteen-year-old girl, embellished her response to a survey question about fear with the comment: "(Nothing scares me!) #Nothing":

36. What scares others usually doesn't scare me. ☐ ☐ ☐ ☑
(Nothing scare's me!) #Nothing

The response of a thirteen-year-old girl with psychopathic traits and serious conduct problems when asked to indicate whether she agreed with the statement "What scares others doesn't scare me." She checked off "applies very well," and so that there would be no mistake about it, added, "(Nothing scare's me!) #Nothing." *Abigail Marsh.*

We didn't find the same pattern for any of the other emotions we asked about, all of which psychopathic and healthy children generally reported experiencing in similar ways. Ours was not the only study to find these effects; several other labs have since produced similar findings confirming that psychopathic children show drastically muted physiological and subjective fear responses.

These findings also beautifully parallel previous findings in S.M., who similarly shows no physiological or subjective fear in response to things that most people find scary. Even attempts to induce extreme fear in her by taking her through a haunted house or handing her pet snakes have yielded no fear response at all—only curiosity. Similar fearlessness has been observed in other patients with severe amygdala lesions, as well as in animals whose amygdalas have been experimentally lesioned. It appears, then, that amygdala damage, whether in association with psychopathy or as a result of Urbach-Wiethe, can result in two unusual and specific impairments: difficulty recognizing others' fear, and a muted personal experience of fear.

To me, this suggested a possibility that goes somewhat beyond what the VIM model and other models of psychopathy propose, which is that amygdala dysfunction in psychopaths impairs not only their behavior but their fundamental ability to empathize with others' fear.

It is widely agreed that an intact amygdala is important for coordinating the array of physiological and subjective processes that result in the experience of fear. This is not the amygdala's only role by a mile, but it is one of its core functions. When an external threat is detected, the sensory cortex conveys detailed information about the nature of that threat to your amygdala: Is it a snake? A gun? The edge of a cliff? The amygdala—which has been described as the most densely interconnected structure in the cerebrum—then rallies the neuronal troops to respond. Messages are conveyed to ancient subcortical brain structures that govern low-level behavioral and hormonal responses to any danger, like the hypothalamus and the brain stem. These structures dutifully ratchet up your heart rate and blood pressure, maximize your air intake, rev up adrenaline production, drive blood into your muscles and away from your core, even pump sugar into your bloodstream for energy. The amygdala also conveys information about the specific threat to various regions of the cortex that enable you to register that a problem has been detected and to alter your ongoing behavior to prevent injury. Without an intact amygdala, none of these processes work very well. The various independent regions all still work, but they cannot be marshaled in the same coordinated way in response to danger.

More, it is thought that the amorphous subjective feeling of fear somehow emerges from the confluence of all this coordinated brain activity, and that too is largely lost in both amygdala lesion patients and highly psychopathic individuals. As one psychopathic sex offender interviewed by the renowned psychopathy researcher Robert Hare responded when asked why he failed to empathize with his victims, "They are frightened, right? But, you see, I don't really understand it. I've been scared myself, and it wasn't unpleasant."

I think we can agree that this is not the statement of someone who really understands what it means to feel fear.

And if someone doesn't understand what it means to feel fear, how can they possibly be expected to empathize with this emotion in others? In fact, as our accumulated data suggested, they can't. Without a normally functioning amygdala, psychopathic adolescents—and presumably adults as well—don't recognize others' fear for what it is, they don't understand how a frightened person is feeling, and they don't, as a consequence, appreciate what is wrong with making someone feel this way. More recent studies I have conducted with my student Elise Cardinale show that, unlike most people, those with psychopathic traits think it is fine to cause others fear by using threats like, "I could easily hurt you," or, "You better watch your back." In an fMRI study, we demonstrated that these aberrant judgments correspond to reduced amygdala activity as these individuals arrive at their judgments.

When Amber hissed threats of arson and violence at her parents, when Dylan held up a knife to his mother and threatened to cut her, when Brianna vowed to beat her schoolmates to a pulp, they did so because they had learned that threatening violence was a useful tool that would help them get their way, but they had no deep appreciation of the emotional suffering these threats caused. Dysfunction in the amygdala and the network of brain regions to which it's connected had robbed them of an essential form of empathy, which is the simple ability to understand another person's experience of fear. They might have had difficulty labeling the emotion their threats caused as "fear," and they almost certainly would not have been able to accurately describe how it felt or truly understand why it was wrong to cause it.

4

THE OTHER SIDE OF THE CURVE

IN 2008, AFTER more than four years at NIMH, I wrapped up my postdoctoral position. The Department of Psychology at Georgetown University, a few miles to the south in Washington, DC, had advertised a tenure-track position for a cognitive neuroscience researcher specializing in social and affective processes, with a focus on child development. It was an enormous stroke of good fortune. The odds of a PhD landing *any* tenure-track professorship are low. To snag one requires luck and good timing—a university where you'd like to work must be looking for someone with your particular research focus and skills the same year that you are looking for a position. Then you just need to beat out another hundred or so applicants for it. That year, both luck and timing were on my side, and I was offered the spot at Georgetown, where I have been ever since. But my research program has expanded considerably to include research with individuals who are as *unlike* psychopaths as you could possibly imagine.

I have not, however, abandoned my brain scanning research with children with serious behavior problems. All told, I have now spent over a decade on this work. And the research my students at Georgetown and I have conducted has expanded our initial findings about the role of the amygdala in understanding others' fear. For example, my student Joana Vieira and others have now found that psychopathic adults and adolescents have amygdalas that are not only underactive but smaller than average as well. In one study by Adrian Raine, the amygdalas of psychopathic adults were found to be about 20 percent smaller than those of controls. Another recent study even found that young men with smaller amygdalas were more likely to have been psychopathic even as children and were more likely to engage in persistent violence in the future. Essi Viding's studies of adolescents have found that the severity of psychopathic personality traits in this age group corresponds to how densely concentrated the gray matter in their amygdala is.

And in a 2014 study of amygdala activity, my student Leah Lozier found still more direct evidence that a lack of amygdala responsiveness to others' fear drives psychopathic children's antisocial behavior. For that study, we scanned the brains of over thirty children with serious conduct problems—children who frequently fought, stole, lied, and broke rules. Some of them had psychopathic traits, like low levels of traits like compassion, caring, and remorsefulness, and others didn't. As I've described, it has long been suspected that antisocial behavior in children with versus without psychopathic traits is driven by distinct brain processes, and this study helped to confirm that. In children who had conduct problems in the absence of psychopathic traits, the amygdala responded *more* strongly to fearful expressions than it did in normal children. This is consistent with the idea that these children's antisocial behavior is a by-product of overreactive emotionality, which may cause them to erupt in response to ambiguous or mild threats. Patterns like this are sometimes linked to anxiety, depression, or exposure to trauma. On the surface, however, it's not always easy to tell these children apart from those who have psychopathic traits, as I had learned early on in my research at the NIH.

Liz and I had recruited a boy named Daniel toward the very beginning of our study—he was perhaps the third or fourth teenager we evaluated. Daniel was unlike most of our other participants in several ways. Nearly all the children I have described so far were white and living in intact, functional families in middle-class or even wealthy neighborhoods, and they went to good schools. Not all the children we studied fit this description, but a lot of them did, and the seeming ease and normalcy of their lives made their cruel and disruptive behavior much harder to explain than if they had led lives of real hardship, as Daniel had.

Daniel was the most outwardly alarming of all the subjects we brought in. Fifteen years old when we first met him, he was a rangy black boy who stood nearly six feet tall and walked with an unhurried saunter and a flat stare. Every time I saw him he was wearing his standard uniform of black sneakers, voluminously baggy black jeans, and a pristine white T-shirt. Some days he topped it off with a black bandanna tied around his head in such a way that his hair poofed out in perfect spheres on either side of his head like Mickey Mouse ears. He once told us about the protracted security screening he always received coming through the NIH gates, and I wasn't remotely surprised. I saw how people responded to him just walking through the NIH corridors. Patients and physicians alike scuttled out of his way like shore crabs whose rock had been overturned, casting furtive sideways glances at him. Walking next to him was an odd experience—as a small woman, I've never personally experienced responses like that in my life. What must it feel like to send ripples of alarm through strangers around you wherever you go? I will never really know, of course, and I never thought to ask Daniel.

The security screeners and NIH personnel weren't wrong to worry about Daniel. He had engaged in more theft, overt violence, and other criminal behavior than nearly any of our other subjects. He couldn't count the number of fights he'd been in. He had been shot at and had shot at other people. He stole from neighbors and stores and restaurants and had tried a variety of drugs. In his telling, he felt neither fear nor remorse about any of it. We had only limited background information about him, unfortunately, because his mother's own mental health problems were severe, so we stuck to a brief and not very informative interview with his aunt. On the basis of this and our interview with Daniel, he received a high enough psychopathy score to qualify for a brain scan.

His MRI session started like any other, with my explaining how the brain scan would work and what he'd need to do while we sat in the cramped control room, which looked like a shabby sort of mission control, with tangles of wires and monitors and knob- and button-bedecked consoles cluttering every surface. As I talked, I noticed that his eyes kept darting toward the control room window, through which the gray, humming hulk of the scanner could be seen. "Something wrong, Daniel?" I eventually asked. "Any questions I can answer for you?"

"What's this going to feel like? Is it . . . it gonna hurt?"

"Oh my gosh, no, definitely not, Daniel! We wouldn't ask you to do something that hurt. It's really just a big camera. Does it hurt to get your picture taken?"

He shook his head.

"Well, this is just the same. It doesn't feel like anything."

He nodded. But I glanced past his head at Liz, who raised her eyebrows at me. The fact that he even asked the question was odd. No other kids with psychopathic traits had asked anything like this. Before the scan, they only ever seemed either mildly curious or bored. Even our healthy controls rarely asked for reassurance.

As the scanning preparations went on, Daniel kept asking more questions. How long would it last? Could we stop the scan if he wanted to come out early? How many other people had we scanned? Could his cousin (who was in the waiting room) come in the scanning room with him? He didn't want to be in the scanner alone. No, we couldn't bring his cousin in the scanning room, but we could bring him into the control room, which we did.

"How you doing, man?" asked his cousin, surveying the scene.

"I'm good," said Daniel. But he didn't look good. He looked nervous.

I opened the door to the magnet. We were ready to start. "Okay, Daniel. We're all set. Think you're ready to go in?"

But Daniel didn't get up. He just stared through the door at the magnet.

Finally, he shook his head. I was astonished to see his eyes welling up. "I can't do it. I can't do it. I want to go. I want my mom."

He wanted his *mom?* He was too *nervous* to go in the MRI? This hardened teenage veteran of gun battles and drug deals was too scared to do something that several sweet ten-year-olds had already sailed through? But it was what happened next that really floored me: he apologized.

"I'm really sorry, guys," he said. "I can't do it. I wanted to do it, though. I thought I could do it."

Then he stood up and grabbed me. He pulled me into a big bear hug.

From inside a nest of skinny teenage arms I managed a muffled "It's totally okay, Daniel. Of course it's okay. Thank you for trying. I'm glad you came in today anyhow."

Daniel had *totally* fooled us. He was a boy who had been forced to adopt the trappings of a hardened adult, and had done so convincingly. But he was not a child with a broken VIM—he was a child whose broken life had pushed him to engage in the same sorts of behaviors that an actually callous, remorseless child would. Underneath, Daniel was, I believe now, probably an ordinary boy in the very best sense, one capable of affection and compassion and remorse who deserved (as do so many children) a much better shake in life than he had gotten. By now he would be about twenty-six years old. I still think of him and his sweet, fierce hug and hope he somehow managed to overcome all the obstacles life had thrown in his path.

What my student Leah's study found is that the brains of violent but emotionally sensitive children like Daniel (at least, the ones who are willing to be scanned) can, on average, be distinguished from those of children who are violent but callous. The Daniels of the world are actually highly emotionally reactive (no matter how well they try to conceal it) and show elevated amygdala responses to others' fear. In contrast, children who are legitimately callous and remorseless show very little amygdala response to others' fear. More, the degree to which actually callous children's amygdalas are nonresponsive to fear seems to be a biomarker of sorts for aggression—particularly the goal-directed, proactive kind of aggression that is notoriously linked to psychopathy. In our study, the relationship between the severity of a child's psychopathic traits and the severity of their proactive aggression could be accounted for, statistically speaking, by how underreactive the child's amygdala was to fearful expressions.

It was powerful support for the idea that the way our brains respond to others' distress is intrinsic to our capacity to experience caring and concern for others.

⁂

I am often asked if doing this kind of research is depressing. Sometimes it is, of course. I feel tremendous sympathy and sorrow for the parents of the children we meet, who are worried and anxious and frustrated, understandably, about their children. I wish I could do more to help them. And I worry for their children's futures. But I really enjoy working with the children themselves. Children who are callous and remorseless are rarely terribly anxious or unhappy themselves—quite the contrary.

To get an overall sense of their mental well-being we sometimes ask children to rate themselves on a scale from 1 to 10, with 1 meaning that they are very unhappy with themselves and 10 meaning that they think they are terrific. A typical child will usually respond with a 7 or 8. But I have heard children with psychopathic traits shout out "Ten!" "Eleven!" even "Twenty!" And remember, these are children who may have been kicked out of multiple schools, who have been arrested, who have no real friends, and whose parents live in fear of them. It's a great reminder of the vast gulf that can separate perception from reality.

The children are often impish and funny and interesting, just like any other teenagers, only more so. Sometimes they can be exasperating—like the psychopathic boy who was getting bored toward the end of his MRI scan here at Georgetown and tried to convince my graduate students to let him come out early because, he said, somehow as he was lying immobile inside the MRI and pushing buttons, his leg had broken. My students had trouble keeping a straight face in response to that one. Other children have caused various kinds of trouble on testing day. One boy locked his mother out of their house when it was time to leave for their scanning session, then refused to let her back in. Another stole food from a cafeteria right outside the scanner. He ate it, unconcerned, in the waiting room. A couple of the girls peed all over their pregnancy testing kits, clearly not bothered that I would be handling them afterward. An enormous number of the boys seem to rarely bathe, judging from the way their feet smelled when they took their shoes off before their scan. But their confidence was untarnished. One memorable teenage boy flirted so incessantly with my flustered twenty-six-year-old graduate student that his mother asked, only half-jokingly, if he was going to invite her to his prom.

But usually by the time they made it all the way into the MRI, the children were motivated to complete their testing and get on to the part where they got some money and a picture of the inside of their brain (the printout of which they could flap in their mother's face, crowing, "See, look, I really have one!").

Unexpectedly, one aspect of this line of research has been downright uplifting, and that is the stark contrast it has revealed between the highly psychopathic adolescents and young adults we have studied and everyone else.

Individuals who are considered "highly psychopathic" make up maybe 1 or 2 percent of the population. This small minority is not categorically different from everyone else, though, as far as we can tell. Rather, they have a larger agglomeration of traits that are present in smaller amounts in much of the rest of the population. One study found that perhaps 30 percent of the population registers as at least a *little* psychopathic on a variant of the PCL used for screening adults in the general community, the Psychopathy Checklist—Screening Version (PCL-SV). Interestingly, that's about the same percentage of volunteers who were unwilling to take any shocks to help "Elaine." It's also about the same percentage of people found by my colleague David Rand to behave uniformly selfishly toward strangers in an online study he conducted. Subjects in that study had the option to share a small stake of money with a stranger they would never meet, for no reason except sheer generosity. Some 39 percent of the subjects never shared the money. But the remaining 61 percent were generous at least some of the time. Likewise, about 70 percent of people would score a big fat zero on a standard measure of psychopathy. That is an incredibly reassuring number.

It's an all-too-common perception that human nature is "fundamentally selfish"—egotistical, Machiavellian, callous. Philosophers have been fretting over this issue for millennia, at least as far back as Aristotle, who concluded that, "All the friendly feelings are derived to others from those that have the Self primarily for their object." Even a person's apparently selfless deeds are, according to Aristotle, ultimately performed to accrue "honor and praise on himself." This line of thinking dictates that even when people *appear* to be acting in caring ways toward others, their behavior always somehow can be traced back to concern for themselves. Did someone give money to charity? A tax write-off! Volunteer to help the homeless? Trying to feel better about his own life! Rescue a woman from a fiery inferno, Cory Booker style, and risk getting burned to death? Well—there must be some self-serving reason buried in there! "Honor and praise," perhaps.

The belief that human nature is fundamentally selfish remains a cornerstone of much modern economic, biological, and psychological research. It's the basis, for example, of the economic assumption of so-called rational self-interest, according to which all human motivation can be reduced to a little internal ledger that calculates the benefits and costs of any potential decision or course of action and strives to pick the option that maximizes benefits for the self—the option that, very simply, is the most selfish. Belief in this view of human nature is pervasive. In 1988, when a representative sample of over 2,000 Americans was asked, "Is the tendency of people to look out for their own interests a serious problem in the United States?" 80 percent agreed that it was. In 1999, a *New York Times*/CBS poll of nearly 1,200 Americans similarly found that 60 percent believed that most people are overly concerned with themselves and not concerned enough about other people, and 63 percent believed that most people cannot be trusted. (A nearly identical percentage of people polled by the 2014 General Social Survey [GSS] agreed that most people cannot be trusted.) Forty-three percent agreed that most people are just looking out for themselves.

But the studies of Milgram, Batson, Blair, and many others, including myself, raise a problem with this view of human nature: people *vary*. There is no one "human nature." To take one obvious example, some people are psychopathic. And if you want to know what a person who is genuinely, fundamentally selfish looks like, just look at a psychopath. They are the Aristotelean ideal of a person whose apparently friendly or helpful actions always really *do* have the self primarily for their object. They are genuinely unmoved by others' suffering and unmotivated to relieve or prevent it. Their apparently good deeds really are aimed at achieving some self-benefit. Take Brent, for instance, a psychopathic boy we studied at NIMH, who fashioned himself as a middle school Robin Hood and sought out bullies to beat up after school—but only to increase his own status and keep others afraid of him and in his debt. The whole point of studying people like Brent—of singling them out and evaluating them with clinical measures—is that they are *different from other people*. Their callousness and indifference toward others' suffering, their willingness to manipulate and exploit others for their own benefit, are not normal. Studying people with psychopathic traits is a great way to gain a renewed appreciation for the fact that most people are not like them at all but instead seem to be genuinely capable of caring about the needs of other people.

Now, saying that most people aren't psychopaths isn't exactly a ringing endorsement of their characters. But the fact that psychopathy is continuously distributed implies something more interesting than just the fact that psychopaths lie at the far end of the callousness spectrum, with most other people bunched up toward the "zero" end. Most human physiological and personality traits aren't distributed in this uneven way. Most traits, from height to cholesterol levels to intelligence to personality traits like extraversion, are distributed in a bell-shaped curve across the population, with most people clustered in the middle of the scale and fewer people residing at either the low or high end. So, for example, the average height of an American woman is about five-four, and the height of about two-thirds of all the women in the country is within a couple inches of that. Only a small number of women are, like me, shorter than five-one; a similarly small percentage are taller than five-seven. Most other variable traits are distributed the same way. This pattern of distribution is so common that it is literally called the *normal curve*.

Psychopathic traits as measured by the Psychopathy Checklist don't fit this distribution. Instead, as one recent study found, they are distributed in what is called a *half normal curve*—it looks like a typical bell-shaped curve that has been sliced in half down the middle, leaving only the right-hand side. What this odd distribution suggests is that psychopathy measures like this are not capturing all of the available variance in traits related to psychopathy, like empathic concern and compassion. Instead, the "half normal" distribution may indeed represent only half of what is actually a symmetrical bell-shaped distribution, which ranges from people with unusually low levels of concern for others at one end (the psychopaths) to a clump of ordinary people in the middle, and then, perhaps, moving past that group, to another, smallish group of individuals on the *opposite* end of the curve from psychopaths whose capacity for caring and compassion is even higher than average—the "anti-psychopaths," you might call them. If this is true—if the small population of truly psychopathic people among us is really balanced out by another small population of anti-psychopaths—it would be compelling evidence that selflessness is just as fundamental to human nature as selfishness.

But who are these anti-psychopaths? Almost no attempts have been made to find them or study them systematically—until now.

The idea to find and study anti-psychopaths may originally have been sparked by, of all things, a paper on face recognition published by some of my former colleagues at Harvard right around the time I

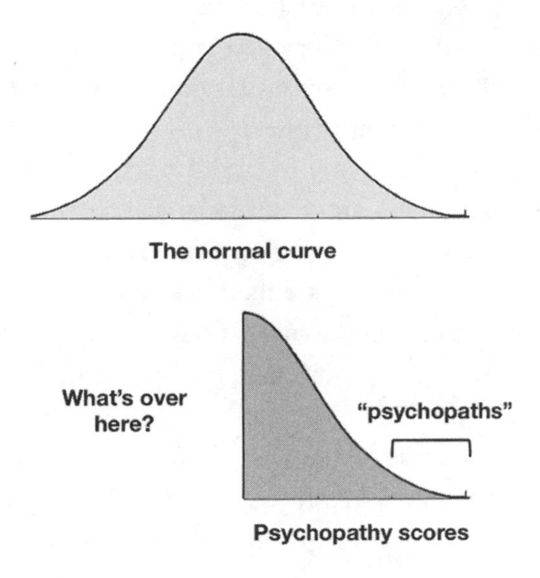

A half normal curve. *Abigail Marsh.*

began working at Georgetown. Recognizing a person's identity from their face is an important and astonishingly complex feat—one we may fail to realize is astonishingly complex because we are so incredibly good at it, better even than the best computers to this day. (The face-sorting algorithms in Google Photos still can't quite keep my children's faces straight.)

At least, most people are incredibly good at it. For over 100 years, rare cases have been documented in which a stroke or head injury has caused a person to suddenly develop *prosopagnosia,* or "face blindness," which all but eliminates the ability to recognize individual faces. Affected individuals no longer recognize the faces of close friends or family members; some even fail to recognize their own face in a mirror. Much more recently, it has been determined that not only can prosopagnosia emerge in the absence of any injury, but it's not even rare: as many as one in forty people may have developmental prosopagnosia, meaning that they have been effectively face-blind their whole lives. Their numbers include the primatologist Jane

Goodall and the late neuropsychologist Oliver Sacks. People with this condition represent the very low end of the continuum of face recognition abilities, which are quite variable; about 60 percent of this variation reflects genetic causes.

If prosopagnosia is starting to sound parallel to psychopathy in some ways—a developmental disorder that is highly heritable and results in significant impairments in roughly 1 to 2 percent of the population and milder impairments in a big chunk of the rest of the population—you're having the same thoughts I did.

But a recent discovery about face processing makes these parallels even more striking: it turns out that people with unusual abilities populate both ends of the face recognition continuum. Those people with developmental prosopagnosia who populate the low end are counterbalanced by a group of *super* face recognizers on the high end—individuals who are extraordinarily good at remembering and recognizing faces. Super face recognizers might smile and greet a woman passing by on the street who happened to be their waitress five years earlier in a restaurant in a different city. Or they might instantly recognize a middle-aged former schoolmate they haven't seen since elementary school thirty years prior. Super face recognizers' abilities are so astounding that they can come off as freakish or creepy. One super recognizer the Harvard research team studied told them, "I do have to pretend that I don't remember [people], however, because it seems like I stalk them, or that they mean more to me than they do, when I recall that we saw each other once walking on campus four years ago in front of the quad!"

We know that inherited variation contributes to extremes in many basic human traits, like intelligence and height. If inherited variation can also contribute to extremes in complex social skills like the ability to recognize faces, resulting in both those with profound impairments and those with extraordinary talents, it is not so difficult to imagine similar patterns emerging for personality traits like caring and compassion as well. In which case, the psychopathic individuals who populate the very low end of the "caring continuum" should be balanced out by another small population of individuals at the very high end of the continuum who are unusually compassionate. Whereas psychopaths are unusually inclined to harm others to benefit themselves, this opposite population may be unusually inclined to risk harm to themselves to benefit others.

Call them the extraordinary altruists.

The world is absolutely bursting with people who have committed moving acts of altruism. They volunteer to help needy animals or children or the mentally ill. They donate money to strangers in faraway cities or countries. They open up their veins to give their blood to the sick or injured. They remove garments from their own bodies to clothe impoverished people in their communities—sometimes in real time. One 2015 online video showed a woman pulling off her own shoes and socks and giving them to a homeless woman whose calloused feet were bare; in another, a young man on the New York subway pulled off his own shirt and hat and carefully helped a shivering and shirtless homeless man into them. Both of these altruists were unaware that they were being filmed at the time. Although such acts are lovely and heartwarming, I would still describe them as everyday altruism rather than extraordinary altruism, for the simple reason that acts of this kind are so wonderfully *ordinary*.

The generally accepted definition of altruism is "a voluntary behavior aimed at benefiting the welfare of another person." Donations to charity, volunteering, donating blood, and helping a stranger all neatly meet this definition, and all these behaviors are squarely in the big central bulge of the compassion curve because they are all (happily) quite common. How common? The 2016 World Giving Index estimates that, in a given month, 2.4 billion people in the world—over half of all the people they surveyed—give some sort of help to a needy stranger. In addition, 1.5 billion donate money to a charity and over 1 billion people volunteer their time. That is a truly staggering amount of help offered to strangers by ordinary people around the globe—*every month.* The United States is, proportionally, among the most giving countries on the planet, ranking number two out of the 140 countries polled. Seventy-three percent of Americans reported that they had helped a stranger in the past month, 46 percent had volunteered their time, and 63 percent had donated money to charity. The amount of money that Americans give to others in a given year is enormous—in 2015 it was $373 billion, at that point an all-time high. The majority of that money ($265 billion) was given by individuals rather than foundations or corporations, and most of it went to secular causes like health and education. The United States' closest competitors in terms of generosity are Myanmar, which is routinely ranked number one on the World Giving Index, as well as Australia, New Zealand, and Sri Lanka, which rounded out the 2016 World Giving Index slots for numbers three through five.

How common are other kinds of ordinary altruism, like blood donations? The World Giving Index doesn't measure blood donations, but the World Health Organization does. The WHO records roughly 108 million blood donations per year around the world. That number includes, according to the American Red Cross, 14 million units of blood collected annually in the United States, which are donated by some 7 million Americans. Because fewer than four in ten Americans are eligible to donate blood, this means that every year nearly one-quarter of all eligible Americans donate blood. This steady stream of donated blood has been given away completely for free since 1974, when the United States moved to an all-volunteer blood supply. In some ways, blood donations are a victim of their own ubiquity. How common they are makes it easy to stop appreciating the generosity of all the millions of people who permit their own blood to be siphoned out of them by strangers so it can be carted away to a blood bank, then injected into still other strangers.

But the risks and discomforts associated with giving blood away to strangers are very minor, which helps to explain why so many people give. The same cannot be said for the donation of other body products, like bone marrow or peripheral stem cells. I'm familiar with the process of stem cell donation, as my mother went through it over a decade ago to save the life of her sister, who was deathly ill from otherwise untreatable lymphoma (and who is, happily, alive and well today). For blood cancer patients like my aunt, such donations often represent the only available cure.

Stem cell and marrow donations involve considerably more time, effort, and discomfort than donating blood. Either type of donation is preceded by hours of medical testing and screening. In addition, for several days leading up to the donation, donors must inject themselves with a medication called filgrastim, which increases the body's production of the life-saving cells but can also cause bone and muscle pain. Finally, peripheral stem cells are harvested from the bloodstream via an hours-long extraction process during which the donor's entire blood supply is siphoned out through a needle in one arm, run through a filter, and returned to their body through a needle in the other arm. So donating stem cells is not exactly a fun process, although it's less intrusive than donating bone marrow, a surgical procedure during which marrow is extracted directly from the bones with a needle. Getting back to normal after either procedure takes donors anywhere from a few days to a month or more. But despite all this—despite the effort and inconvenience and discomfort and the fact that they cannot receive any payment at all—an enormous number of people have donated marrow or peripheral stem cells to save the lives of total strangers. Some 10 million people are registered as potential donors in the National Bone Marrow Registry, and about 5,000 of them donate each year.

Donating marrow or stem cells to a stranger begins to strain the limits of what can reasonably be termed "everyday altruism." That said, bone marrow and peripheral stem cells grow back. Healthy bodies produce them constantly. Removing them is very low-risk and causes only temporary discomfort. For this reason, although donating marrow or stem cells to a stranger is a wonderful and admirable thing to do, I still would not term it truly *extraordinary* altruism.

Truly extraordinary altruism must not only meet all the usual benchmarks of altruism—a voluntary behavior aimed at benefiting someone other than the self—but it should go beyond ordinary altruism in three ways. First, the beneficiary should be someone unrelated and unknown to the altruist at the time they decided to act. Second, the act must present a significant risk or cost to the altruist. Third, the behavior should be non-normative—something people are not expected or taught to do. Not only does an act that satisfies all of these stipulations impress us as morally exceptional, but it becomes very difficult—impossible, arguably—to attribute the cause of the act to anything but genuinely altruistic motivation, because no amount of fiddling with the ledger could make the benefits to the altruist exceed the costs.

This is important to establish for two reasons. First, truly extraordinary altruism, a behavior that lies at the far end of the caring continuum, is most likely to represent a true mirror image of psychopathy, which lies at the other end. An extraordinary altruist is the most likely to genuinely show us what an anti-psychopath might look like. Exploring extraordinary altruism is also important because a sizable subset of the population is resistant to the possibility of truly altruistic motivation. And I have learned that even people who are receptive to the idea of altruism in the abstract often suspect that particular instances of apparent altruism are actually driven by selfish motives. To have any hope of identifying the neural underpinnings of the human capacity for altruism, it is important to first find examples of it that we can generally agree are driven by genuinely altruistic motivation rather than some other more self-serving cause.

What makes things tricky is that most human behavior is multiply motivated—driven by many forces operating on many levels, both conscious and unconscious, and many of these forces are indeed self-serving. Take an example of ordinary altruistic behavior from my own recent past: I invited my younger brother to live in my basement for a month while he looked for a new home for his family. At a behavioral level, this was surely altruistic: it was a voluntary offer on my part that entailed some (small) costs to myself, and the offer was aimed at benefiting my brother's welfare—saving him some money, keeping him fed, making sure he was comfortable. But a skeptic could ask, "Was it *really* altruistic? Was the behavior really *primarily* motivated by your desire to help him?"

The answer isn't as easy as me retorting, "Of course it was! I'm the one inside my own brain, and I *know* the reason I helped my brother, and it was because I wanted to improve his welfare."

This response assumes that I can know the reasons for my own decisions and behavior. But this assumption is baseless, in part because the "me" that I am aware of isn't really inhabiting my entire brain. The conscious mind is aware of only a tiny fraction of the processes going on inside the whole brain. What is your pituitary up to right now? How about your brain stem? Are you aware of them? Can you control them? Do you know *why* they are doing what they're doing? No. They are more or less cut off from your conscious mind, as are many other neural processes. As a result, although people are reasonably good at correctly reporting on things that they can observe directly, like their own behavior, they are only so-so at reporting internal states like how they are feeling, and they often have no idea *why* they are feeling or acting as they are—and can be easily misled.

There are countless demonstrations of this in the psychology literature. One I like comes from one of Daniel Batson's studies of altruism in which he administered a pill called Millentana to his participants at the outset of the study. As he gave them the pill he told some of the participants that Millentana was a drug that would make them feel very warm and sensitive. Others he told that it would make them feel uneasy and uncomfortable. Both explanations were pure fiction—Millentana was actually cornstarch, a psychologically inert placebo.

After swallowing the Millentana and a fictitious explanation of its effects, participants watched a stranger receiving painful shocks. After the first few, some participants were given the option to either receive the remainder of the shocks in her place or leave the study early (with the awareness that the stranger would keep receiving shocks after they departed). Batson found that 83 percent of those who believed that Millentana was making them feel warm and sensitive volunteered to take the shocks for the stranger whereas only 33 percent of the uneasy and distressed participants did. What produced this massive difference in behavior? A little trick that duped the participants into believing that a pellet of cornstarch was fiddling with their emotional states and massively changing their tendency to act altruistically. There is no record of Batson asking the "warm and sensitive" participants *why* they felt and acted more than twice as altruistic as the "uneasy" participants, but they presumably would have chalked it up to the drug, which Batson knew (as do we) actually had no effect. Batson's experiment was a stark demonstration of how easily feelings, motivations, and behaviors can be shaped by forces outside of people's awareness.

Two forces that biologists have convincingly shown to be drivers of everyday altruistic behavior—and that could well have played an unconscious role in motivating my offer to help my brother—are *inclusive fitness,* which biases organisms to help genetic relatives, and *reciprocal altruism,* which biases organisms to help those they interact with frequently, related or not. Both of these forces are ultimately self-serving, although the behaviors they promote are no less beneficial for others. Inclusive fitness drives altruism toward relatives across many species and promotes successful propagation of the altruist's own genes. The thinking is that by helping my brother, who shares about 50 percent of my genes, I'm helping my own genetic legacy in some small way. Any help I give him improves his fitness and thus his likelihood of passing along his own genes—and by extension mine. It's a convincing explanation for why the preponderance of costly helping behavior across species, from ants to birds to people, benefits genetic relatives. This drive to help genetic relatives occurs somewhere deep down in the nervous system via mechanisms that we share with monkeys and birds and ants, and may not require our conscious awareness. So I can't possibly know to what extent this phenomenon affected my offer to temporarily house my brother—although I would readily admit that I would have been unlikely to make a similar offer to, say, a fourth cousin.

"Aha!" you might say, "but we help family not just because we are closely related to them genetically, but because we are emotionally close to them." This is true, and this is where reciprocal altruism comes in. We offer most of our altruistic helping to people with whom we have long-standing close relationships or with whom we share membership in an important social group, whether it be a family, a neighborhood, a workplace, or a group of friends. The rules of reciprocal altruism are simple: help those who have helped you in the past or who are likely to help you in the future. I bring you some coffee today, you spot me money for lunch tomorrow. I help raise your barn today, you scare a thief away from my cattle tomorrow. I help you escape danger today, and maybe you'll help me out of a pinch tomorrow or at some other future point. When everyone follows the rules of reciprocal altruism, which members of social species largely do, everyone prospers. It's a game with very good long-term odds. The whole setup only works, though, when applied to those with whom you expect to keep interacting and who can therefore be reasonably expected to reciprocate. When people interact with strangers they have no expectation of ever seeing again, altruism, especially costly altruism, tends to plummet.

Given this, it's amazing in some ways that we are *still* as nice as we are to strangers—giving them directions and making change and holding doors open and giving them donations of money and blood. Neither inclusive fitness nor reciprocity can explain such behaviors. Often, these various forms of low-cost helping may indeed be driven primarily by a selfless desire to help others, although still other forces can be at play as well. For one, because these kinds of behavior satisfy established social norms, we often engage in them merely from habit or by default and, as a bonus, get to feel a sense of satisfaction of having lived up to cultural ideals (whereas we might feel shame if we failed to do so). And sometimes these low-cost behaviors personally benefit the altruist in ways that go beyond future reciprocity. I benefited, for example, from helping my brother. He is fun and funny and interesting, and I loved getting to spend extra time with him while he was staying with me. Likewise, charitable donations can result in tax write-offs, and gallantly holding open doors and volunteering for charities can yield valuable boosts to social reputation. Any self-serving benefits may be small for a given instance of these forms of altruism, but because the costs are also low, the benefits still outweigh them. Thus, although all of these kinds of low-cost altruistic behaviors are wonderful—and often at least partly motivated by altruism—it's nearly impossible to say for sure what is driving any single instance of donating or direction-giving or change-making or door-holding or basement-letting because there are so many possibilities, and because the various reasons can operate in tandem—a single instance of giving can simultaneously reflect kin selection, expectations of reciprocity, social norms, and a genuine desire to help.

This brings us back to the question of extraordinary altruism, which, again, is a voluntary behavior aimed at benefiting a stranger *and* which is non-normative *and* presents a serious risk or cost to the altruist. These stipulations ensure that, in contrast to most acts of ordinary altruism, all the possible alternative motivations have more or less been ruled out. The most common alternative possibilities—that the act was primarily motivated by kin selection, expectations of reciprocity, social norms, habit, or self-benefit—have all been stripped away. The stringency of this definition means that very few behaviors unambiguously qualify. One that does is the heroic rescue of a stranger—like the rescue that saved my life back in 1996.

When it first occurred to me that studying people who are unusually caring might be just as informative as studying those who are unusually uncaring, my thoughts naturally turned to heroic rescuers. I thought first of Lenny Skutnik, one of the more famous names in altruism research. Skutnik was a Congressional Budget Office employee in Washington, DC, who was carpooling home to Lorton, Virginia, one frigid January afternoon in 1982 when an insufficiently de-iced plane took off from National Airport, stalled out, lost altitude, then plummeted into the Potomac River near where Skutnik's car was idling in traffic. Helicopters arrived some twenty minutes later to retrieve the few surviving passengers from the ice-choked river. One of them, Priscilla Tirado, was by then so weakened by hypothermia and panic that she slipped from a rescue line back into the water as Skutnik raced to a nearby riverbank. When he arrived, the scene was eerily quiet. Then a woman's terrified voice pierced the quiet: "Will someone please help!?"

Skutnik's immediate reaction—"like a bolt of lightning or something hit me," he later said—was to strip off his coat and boots and hurl himself into the 29-degree water. He swam out some thirty feet to retrieve the half-frozen Tirado and haul her back to safety. It was an incredibly dangerous thing to do. Another would-be rescuer had already been forced back by the punishing cold and ice floes. Even back on shore, Skutnik's focus remained on the others around him. As he sat shivering in an ambulance that had run short on blankets, he gave another soaked survivor his own coat. For his deeds, Skutnik received a Carnegie Hero Fund Medal and an invitation from President Reagan to attend a State of the Union Address, at which he was hailed as an American hero—albeit a reluctant one. Like Cory Booker, he consistently resisted being labeled a hero and was never comfortable with the adulation and attention he received.

I was tempted to reach out to Skutnik. He is a canonical example of real-world extraordinary altruism who I knew had lived in northern Virginia, minutes from my office in Georgetown. Assuming he hadn't moved, I could probably walk to his house. But, frustratingly, for the purposes of trying to contact him, he might as well have lived on Venus.

As a university scientist who studies human behavior, I am bound by the rules of an institutional review board (IRB), the duty of which is to protect the welfare of university research participants. My research subjects don't need a lot of protecting, as my studies are not terribly risky. I'm not allowed to call filling out surveys or participating in a brain scan "no risk," as it's technically possible to suffer some sort of harm anywhere. A subject could get a paper cut from a questionnaire, or experience claustrophobia inside the MRI scanner (and of course there's the risk of serious harm if metal is introduced). But these research techniques are considered "minimally risky"—that is, they are no more risky than other routine activities like going to school or seeing a doctor. But risks and benefits must be considered in relation to one another, and my research also does not benefit my participants at all. They aren't receiving treatment or therapy or training that might help them personally. So to ensure a favorable risk-benefit ratio, I am required to be cautious about avoiding any practice that might make participants feel pressured to subject themselves to even the low risks that my research presents. So, for example, I can't pay them too much. I can't offer to pay a fourteen-year-old $1,000 for a half-hour brain scan. The chance to earn that much money might make even a very claustrophobic child feel like he had no choice but to take part. I also am required to use no-pressure recruitment tactics. I can post advertisements on flyers or in newspapers or email listservs because those advertisements don't leave anyone feeling personally obligated to respond. What I absolutely could not do under any circumstances was cold-call Lenny Skutnik to ask him if he'd like to take part in a brain imaging study. Although a journalist or writer or market researcher or second-grader doing a class project or literally anyone *other than a university-affiliated researcher* could legally and ethically look up Skutnik's name in the directory and give him a call if they wanted to ask him some questions, I, as a university-affiliated researcher, could not. And that was that.

Luckily, heroic rescuers aren't the only people out there who meet the requirements of extraordinary altruism. Some two decades ago, another form of extraordinary altruism was born, one that has been described as the moral equivalent of saving a drowning stranger: altruistic organ donation, which is the donation of an internal organ, usually a kidney, to a stranger. In stark opposition to a psychopath like Gary Ridgway, who destroyed a stranger's kidney in an attempt to end his life, these altruists give a stranger a kidney in an effort to save someone else's life.

This kind of donation is quite a recent phenomenon. Before the 1990s, donating a kidney to anyone who wasn't a relative was considered an "impenetrable taboo." Most transplant physicians would refuse to perform the surgery. The reason had little to do with the technical difficulty of transplanting an organ between strangers. The first successful kidney transplant from a living donor was recorded in 1954, and the first from a living donor who was genetically unrelated to the recipient in 1967. It also certainly had nothing to do with a lack of need. Then, as now, the number of people with end-stage renal disease who desperately needed a kidney and couldn't find either a deceased or living donor grew every year. Today over 90,000 people are on the wait list. So why did it take so long until most transplant centers would consider transplants from altruistic donors?

The reason largely boils down to the pernicious belief that human nature is fundamentally selfish.

Unlike completing a questionnaire or an MRI, donating an organ entails real risks. Surgeons know this better than anyone. Physicians' first and most important oath is *primum non nocere:* first, do no harm. A successful surgery is one after which the patient wakes up better off than they were before, or at least no worse off, an outcome that requires a team of surgeons, nurses, technicians, and anesthesiologists to carry out dozens of delicate and precise maneuvers exactly right. Even when they do, unforeseeable mishaps sometimes occur. Infections, clots, and bad reactions to anesthesia are just a few of the complications that can cause surgery to go south. These issues are rare, thankfully, during modern kidney removals: only about one in 50 nephrectomies results in serious complications like bleeding, and only one in 3,000 results in death.

So donating a kidney is actually considered a low-risk surgery. But to put it in perspective, compare the risk of kidney donation to the risk of skydiving—that pursuit of risk-takers and adrenaline junkies. The odds of dying after tumbling out of a plane beneath a parachute are about one in 100,000, which means that the risk of death from donating a kidney is over *thirty times* higher. And unlike skydiving, kidney donations pose some long-term risks as well. Officially, living kidney donors have about the same long-term health outcomes as the average person. But donors must be *healthier* than average to qualify for surgery. High blood pressure, obesity, or diabetes all rule out donation. So if people who start out with above-average health have only average lifetime outcomes following surgery, it suggests that the loss of a kidney could entail slight long-term risks, like increased blood pressure and a risk of kidney failure.

But what really separates kidney donation from surgeries that are not considered "impenetrably taboo" is not its risks—which, again, are not high—but its complete lack of benefits for the donor, at least from a medical perspective. Comparably risky surgeries are usually performed to remove organs that are diseased or causing the patient pain or even just inconvenience, like the removal of a gallbladder prone to stones or of a uterus to prevent pregnancy. And of course, millions of non-zero-risk surgeries have been performed for decades for purely cosmetic reasons. What makes these surgeries ethically uncomplicated, though, is that all their risks and benefits, however minor, redound to the same person. This person has presumably decided that the balance between risks and benefits is favorable, and that surgery is ultimately in his or her best interest.

What makes a kidney donation different is not its overall risk-benefit ratio, which is very favorable. It is that the risks and benefits are shared—unequally—between two people. The donor volunteers to take on only medical risks to give the recipient all the medical benefits. Living organ donations represent, in the words of the surgeon Dr. Francis Moore, "the first time in the history of medicine [that] a procedure is being adopted in which a perfectly healthy person is injured permanently in order to improve the well-being of another." If you assume that human nature is fundamentally and uniformly selfish, and that all human decisions and behavior "have the Self primarily for their object," the whole thing simply makes no sense.

Nevertheless, after the first successful living kidney transplant, surgeons gradually began performing more such surgeries in the ensuing decades. The problem of kidney failure was not going away—indeed, it kept getting worse. The wait list ballooned with every passing year. But for the most part the only acceptable donors were deemed to be people very closely related to the patient, preferably blood relatives, although in some cases spouses or other relatives would be considered. A few transplant surgeons would even perform donations between unrelated but emotionally close donor-recipient pairs, but most would not. Why? Again, the rationale boiled down to the belief that such a surgery was only justifiable if donors personally stood to benefit at least as much as they risked from the surgery—a rationale based wholly on the norm of self-interest. The thinking went: perhaps a mother who donates a kidney to her daughter, or a husband who donates to his wife, isn't medically benefiting from the procedure. But they will benefit by being spared the grief of losing a loved one, or the hardship of losing someone they depend on, or from having to support the patient through endless rounds of dialysis. Perhaps these benefits, all added together, could outweigh the risks of the surgery. Surgeons would even go so far as to add improved self-esteem to the "gain" side of the donor's ledger. But the idea of removing someone's kidney in the absence of *any* concrete compensatory gains they might accrue by donating remained unfathomable.

What changed in the late 1990s? Arguably, it was (in part) the persistence and openness of a woman who has chosen, until now, to remain anonymous. I can reveal here, with her permission, that her name is Sunyana Graef. She is a sixty-eight-year-old mother of two who lives in Vermont, where she has worked for twenty-eight years as a Zen Buddhist priest. She is one of altruistic kidney donation's "index cases"—a patient whose altruistic donation played a major role in changing the donation landscape. Graef was not the first person ever to donate a kidney to a stranger; there are reports of such donations as early as the 1960s. One detailed case study of another altruistic donor who falsely reported that she knew her recipient (and was later found out) was reported in 1998.

But Graef's was the first reported *nondirected* donation. This is considered the most extreme form of altruistic donation, as the donor does not specify a recipient, does not know the recipient's identity before surgery, and in some cases never learns who received their kidney. A donation of this kind—one that restores an unknown and unspecified receiver to full health and life—achieves the very highest moral status. The Hebrew philosopher Maimonides considered a gift that leaves both the giver and the beneficiary anonymous and that ends the beneficiary's need for further charity to be the very highest form of giving. Ancient Greek philosophers would have considered such an act to exemplify the highest form of love, which they termed *agape*—unconditional love for any person, regardless of circumstances, rather than for any particular person or group. Most relevant for Graef, Zen Buddhism also advocates love and compassion that flows toward all beings, rather than being directed at any one individual.

A second major difference was that, prior to Graef's donation, donations between strangers had largely been performed quietly and, on the part of the surgeons involved, often reluctantly. These donations were rarely formally recorded, and they generated no major cultural shifts. The same cannot be said for Graef's donation.

It was 1998 when Graef first contacted a nationally renowned transplant center in Massachusetts and told them that she had given it some thought and really wanted to give one of her kidneys to someone on the transplant waiting list. She had never heard of anyone giving a kidney to an unspecified stranger before, but it seemed like it should be possible, and such a donation would be in keeping with her Buddhist vow to help all living things. She already volunteered some money and time to help others, but as a mother and full-time priest, she didn't have very much of either, and it didn't feel like enough. But she did have two kidneys. She had read up on, and was comfortable with, the level of risk the surgery would involve, as was her husband. The recipient could be anyone, she said, who didn't kill for a living (like a hunter). And she wanted the donation to be anonymous. Her plan was to register at the hospital under an assumed name and to never meet the person who received her kidney so as not to cause the recipient to feel any debt or obligation to her. What do you think the transplant coordinator said? The response was polite, but flat: No way. Under no circumstances would they entertain the idea of this donation.

Think of it! That year some 35,000 Americans were stuck on the kidney transplant waiting list, most of them too sick to survive for more than a few years without a donor. And as any transplant professional would know, most of them wouldn't find one. Many patients have no family members eligible to donate, and there are never enough deceased donors to make up the difference. (Transplants from deceased donors are less effective anyhow.) And here came a woman offering to give one of these patients a golden ticket—restoration to health and a normal life—by undergoing a surgery that is sufficiently safe that if she had been the patient's sibling or parent it would have been perfectly acceptable. But the transplant center told her no, absolutely not—not because what she was seeking was impossible or even terribly difficult from a surgical perspective, but because what she sought seemed *psychologically* impossible.

For anyone who believes that human nature is fundamentally selfish, Graef's request could only be explained by one of two unappealing alternatives. The first was that her wish reflected some self-interested calculus and she expected to receive benefits sufficient to compensate her for the risks she would assume. But the nature of her request eliminated any concrete benefits. By requesting that the transplant center pick the recipient, she ensured that this person would not be a relative or friend of hers, so her wishes couldn't have been driven by the desire to help a blood relative or by expectations of reciprocity. And her request for anonymity ensured that she couldn't receive any social or financial reward. Organ donors can't legally receive payment anyhow. (This requirement, by the way, makes them the *only* ones involved who get no concrete benefits from their donation—the physicians, technicians, and hospital staff all get paid, and the recipient gets a kidney.) And by never meeting the recipient, Graef wouldn't even have the pleasure of seeing this person returned to good health or hearing the words "thank you."

The only other alternative—again, for anyone who believes that rational self-interest drives all human decisions—was that her wish did *not* reflect a rational calculus. In other words, she was crazy. She was irrational or delusional. Perhaps she believed that undergoing the surgery would fix some problem in her own life. Perhaps she was suicidal and hoping the surgery would go awry. Or perhaps she was seeking medical attention for pathological reasons; she could have been exhibiting a symptom of a rare factitious disorder called Munchausen syndrome. Any of these motivations would render her an unacceptable candidate for surgery.

Fortunately, Graef wasn't content with the first answer she got. In her mind, the donation was already a foregone conclusion at that point. "It was like it wasn't my kidney anymore. I just needed to find a way to make the donation happen," she later recalled to me.

So she next reached out to the kidney transplant program at Brown University, which was (and still is) run by transplant director Dr. Reginald Gohh. To her relief, Dr. Gohh didn't say no. Not that he immediately said yes either. He directed a major transplant center, but he had never heard of a donation request like this before. So Gohh first set up an interview in an attempt to figure out who this unusual woman was and to understand her request better. He came away impressed by her level of knowledge about the transplant and the seeming sincerity of her request. But before Graef could even begin the medical workup that precedes a kidney donation, Gohh wanted her to first speak to an entire team of transplant professionals to see if they also believed her to be both sincere and rational. The team included the transplant coordinator, a social worker, a transplant nephrologist, and a transplant surgeon.

All came away convinced of something that was at that time fairly radical: that this prospective donor wasn't crazy or irrational or deluded, that she was sincerely motivated by altruism, and that, moreover, her motivation for donating was morally admirable and a legitimate reason for proceeding.

The transplant took place on February 8, 1999. A surgical team made a single incision in Graef's abdomen, removed her left kidney, then quickly transferred it to a second operating room to be stitched into the abdomen of the recipient. Both Graef and her recipient—whom she has never met—experienced smooth and uncomplicated recoveries and soon resumed their normal lives. Graef was back at work in her temple a week after the transplant.

But neither Graef nor Dr. Gohh was content to leave it at that. Gohh came away convinced that donations like this one were ethically justified and should be performed when medically appropriate. Graef agreed that, as long as her anonymity would be preserved, it was important to let others know that surgeries like hers were possible. So in early 2000, Dr. Gohh wrote up Graef's case, and the article was published the following year in three brief pages of the medical journal *Nephrology Dialysis Transplantation,* thereby helping to usher in a new era of altruism.

In 1999, the United Network on Organ Sharing (UNOS) recorded five anonymous altruistic kidney donations in the United States. In 2000, there were another twenty, and by 2001 there were thirty more. The numbers increased every year until they reached their peak in 2010, when 205 people anonymously donated their kidneys to strangers. Currently, between 100 and 200 altruistic donations take place in the United States every year. And this number counts only those that, like Graef's, are "nondirected," meaning the transplant center selects the recipient, who usually does not meet the donor before the transplant (although they often meet afterward). Many more donors elect to give their kidney to specific strangers whose need they learn about on Facebook or Reddit or a billboard or through a website like matchingdonors.com. Nearly all transplant centers will now consent to perform either type of altruistic donation, and gone are the days when surgeons would use terms like "repugnant," "offends the human conscience," and "pathologic by psychiatric criteria" (really) to describe these donors. Thousands of lives have been saved by the growing acceptance that a genuine desire to help another person, despite the costs to oneself, may motivate the donation of a kidney.

In 2009, I read "The Kindest Cut," Larissa MacFarquhar's wonderful *New Yorker* article about advances in altruistic kidney donation. It prompted me to do some more background research, which made the many similarities between altruistic kidney donors and other altruists clear. Heroic rescuers tend to make their decisions to help rapidly and intuitively. My colleague David Rand, a behavioral scientist at Yale, has conducted research showing that recipients of the Carnegie Hero Fund Medal overwhelmingly report that their decisions to rescue strangers were fast and spontaneous rather than deliberative. Altruistic kidney donors like Graef tend to report having a similar experience, often stating that when they first realized that they could donate a kidney to a stranger, they just "knew"—as though a bolt of lightning had struck them—that they wanted to do it, and they rarely felt any ambivalence or hesitation afterward. Also, like Cory Booker and Lenny Skutnik and many other heroic rescuers, they tend to be humble about their actions afterward, actively resisting being labeled heroes. Graef maintains to this day that she was really just a "conduit" for the donation, that Dr. Gohh and the surgeons, physicians, nurses, and even secretaries and janitorial staff at Brown's transplant center made the donation possible and were the real donors.*

Altruistic kidney donors are unlike heroic rescuers, however, in one key way as far as research is concerned: they can be contacted en masse, without cold calls or coercion, through transplant centers and listservs. I decided that altruistic kidney donors were the extraordinary altruists who could help me explore the idea of a compassion continuum—the anti-psychopaths whose brains might reveal the roots of human altruism.

I spent the next year seeking funding for the project, which was no small task. Few organizations that fund scientific research have missions and funds compatible with studying the brains of extraordinary altruists. I caught a very lucky break, though. In late 2009, the renowned social psychologist Martin Seligman, in cooperation with the John Templeton Foundation, put out a call for neuroscience research proposals aimed at testing positive features of human nature like morality, resilience, and altruism. Bingo. I applied, and in 2010 I received a $180,000 "Positive Neuroscience" award to conduct the first-ever research on the neural basis of extraordinary altruism.

I originally thought the hardest part of the research might be locating enough altruistic kidney donors to complete the study. I wanted to find twenty altruists whose brains I could scan, and I had only a very small population from which to draw—there were roughly 1,000 nondirected kidney donations ever recorded at that time in the United States, and an unknown number of other altruistic donors. I was willing to fly the donors in from anywhere (and thanks to my grant, I had the money to do so), but who knew how

---

*As Graef describes it, this sentiment aligns with the Buddhist teaching that the highest form of giving is one that recognizes no giver, no receiver, and no gift, and that is derived from an understanding of our inherent oneness.

many eligible volunteers I would be able to find? I would have to rule out anyone with magnetic metal inside their body, for example, and metal clamps are sometimes used in nephrectomies. I couldn't include anyone taking medications for anxiety or depression or chronic pain, or anyone with claustrophobia. And of however many altruists remained who met all our criteria, how many would even want to participate? Normally, recruiting anyone older than college students for psychology research is like pulling teeth. Research doesn't pay enough to attract the average busy, working adult; IRB restrictions against coercion prevent us from paying anyone with a decent income enough to actually compensate them for their time. And that's by design. The goal is for research volunteers to partake in studies, not for the money, but out of a desire to help science and the public at large—out of altruism.

So as you might be able to guess, I didn't actually need to worry about finding study participants at all.

I've actually never experienced anything like it. In early 2011, shortly before I left for a psychology conference in Texas, I had reached out to several organizations that work with kidney donors. I posted recruitment advertisements in a couple of national listservs for living kidney donors, and I asked the Washington Regional Transplant Center to contact the dozen or so altruistic donors in the area whom they had on file.

I didn't have a smart phone back then, so I couldn't check my email until midway through the first day of the conference. When I opened up my laptop and logged in, my jaw literally dropped open. It was like I had tapped into the matrix. My inbox was full of messages from altruistic kidney donors:

> "I donated a kidney to a stranger last February. I would be happy to participate in your study."
>
> "Hi, I saw on Facebook that you are studying living kidney donors. I donated to a stranger in 2009 and would love to participate."
>
> "I am responding to a Facebook post by Abigail Marsh seeking volunteers to participate in your study."

"Please know that I am EXTREMELY interested in participating in the study should the study want me."

"I was an altruistic kidney donor . . . would be more than happy, and honored, to take part in a study."

My favorite may have been:

"I WOULD VERY MUCH BE INTERESTED IN BEING A LAB RAT AND HAVING STUDIES DONE ON ME."

This doesn't happen in the normal world of behavioral research. Maybe it does for researchers who are studying life-saving cancer treatments or paying people thousands of dollars to sleep in a sleep lab, but for basic research on human behavior—no. I had been conducting psychology research for over a decade, and recruitment had invariably been a long, slow slog just to find a sufficient sample of somewhat enthusiastic volunteers. When you're seeking a small, select population, it is even harder. Recruiting and screening twelve eligible adolescents with psychopathic traits for my first fMRI study at NIMH took about two years, and they're not even very rare.

But despite altruistic kidney donors comprising less than 0.001 percent of the population, it took less than two days to recruit twelve of them, and within a week we'd heard from enough altruists to fill the study. The emails they sent were some of the friendliest, most effusive messages I'd ever received from strangers.

It was an apt introduction to the world of extraordinary altruism.

5

WHAT MAKES AN ALTRUIST?

BEFORE WE DELVE deeper into the world of extraordinary altruism, I'd first like to invite you to accompany me on a trip that may be essential to understanding this world. Come ride with me, if you will, on a beam of light. This is not Einstein's beam of light. It will not provide any insights into the conjoining of time and space. This beam will perform a feat, however, that I think is equally spectacular: enabling the conjoining of two human minds.

How information inside one person's head ever makes its way into the interior of someone else's head remains one of the great mysteries of psychology. Language obviously plays a crucial role. It would be nearly impossible to understand cognitively complex phenomena in others, like beliefs and desires and intentions, without language. Think of how much information you can glean about another person's beliefs and goals when they say something like, "Hey, let me try!" or, "I'll do that for you."

But language is a foggy window into the mind. Most internal states are never verbalized. Some thoughts are too private or too banal to express. Other internal states cannot be put into words because they are too complicated, or because they are unknown even to the person experiencing them, lurking somewhere inaccessible in the unconscious. And language can mislead—sometimes intentionally, as in the case of irony or deception, and sometimes simply by happenstance. Is a person who says, "I'll do that for you," being helpful, impatient, or chauvinistic? The words themselves are silent.

Because the stream of spoken speech is only ever a fragmented reflection of the mind that produces it, much of what we know about others' complex mental states—their beliefs and desires and intentions, sometimes called "cold cognitions"—is simply educated guesswork. Sometimes it *seems* like we actually know what intentions underlie an utterance like "I'll do that for you" if it comes from a good friend, or if it's accompanied by a smile versus a sigh. But this so-called knowledge is an illusion. We have no direct access to others' thoughts. The best we can do is patch together inferences about what the people around us believe or intend or want by pinning observations of their behavior into the complex web of knowledge we have about them individually and about people in general. Although most adult human brains can do all of this quite quickly, it's a terrifically complicated process, and it's a wonder we ever get it right. Often we don't, of course. Although most people believe that they are good at understanding others' true internal states, psychological studies of lie detection suggest otherwise. When put to the test, nearly everyone's ability to tell the difference between what others say and what they actually think when the two things differ is little better than chance. You might as well flip a coin.

The same is not true for understanding others' emotions ("hot cognitions"). Although sometimes we infer how other people are feeling using similar inferential processes, that's far from our only option. Actual, valid information about people's internal emotional states is literally pouring out of them all the time in forms that we can access directly with our eyes, ears, hands, and even noses. Clues about others' emotions seep from their pores in chemical form, enabling us to literally smell others' fear. (This is not a myth! It actually happens.) Internal emotional states echo in the pitch and timbre of people's voices, shift the movements and postures and even temperatures of their bodies, and festoon the surface of their faces. This last source of information is particularly important for humans. Our species pays more attention to, puts more weight on, and gets more information from facial movements than any other single channel of information.

Few researchers have contributed as much to figuring out how we use facial movements to understand others' internal states as the psychologists Paul Ekman and Wallace Friesen. In 1978, Ekman and Friesen created the first comprehensive inventory of all possible expressive movements a human face could make. They paid particular attention to facial movements that, when combined, yield one of six widely recognizable emotional expressions: anger, disgust, happiness, sadness, surprise, and—of particular relevance to altruism—fear. They generated a set of black-and-white photographs of these expressions that have been used in thousands of psychology and neuroscience studies around the world in the ensuing decades. Although many other sets of emotional facial expressions have been created since then, all of which offer different strengths, none of them, in my view, has been more carefully constructed and standardized than Ekman and Friesen's originals. The common use of these expressions by investigators around the world has permitted studies of emotion to be replicated by multiple research groups; this is a rare and valuable phenomenon in behavioral research and one that provides a much higher degree of confidence in the research results.

I have used this set of facial expressions myself since I was an undergraduate. In the process, I became very familiar with what Ekman and Friesen look like, as they lent their own faces to their research. I nearly jumped out of my skin the first time I turned around at a conference to see the long, mournful face of Wallace Friesen staring back at me, both familiar and disconcertingly new, in the way that a face only seen before on a screen looks in three dimensions. I imagine he gets that response a lot.

Ekman and Friesen determined that for a face at rest to transform into one that appears fearful, three specific sets of movements are required. First, and most importantly, the *levator* muscles of the upper eyelids must be contracted to pull back the lids and widen the eyes. Human eyes are ideally designed to make this objectively subtle muscle movement obvious. Nearly unique among all species with eyes are the vivid white sclera that surround the human iris. (This is why one tactic animators use to make animals appear more human is to give them bold white sclera. It's a trick you see in movies from *Bambi* to *Finding Nemo* to *The Planet of the Apes*. Actual deer and fish and chimpanzee sclera are dark or hidden. But if you make them larger and paler, the animal suddenly appears human.) The bold visual contrast created by the juxtaposition of white sclera, pigmented skin and iris, and black pupil draws in the viewer's gaze. This effect is made more potent when fear sweeps the lids backward to reveal even more of the gleaming sclera underneath. "Look at me! Meet my gaze!" these sclera scream.

But although wide eyes may draw the viewer in and create an appearance of vulnerability, in isolation they don't yet convey fear. Fearful brows must also contort into a new shape. The *frontalis* muscles in the forehead draw the brows upward toward the hairline, while other muscles like the *corrugator supercilii,* small triangular muscles that overlie the inner corner of the brows, simultaneously crinkle the inner edges of the brows inward and slightly down. (The goal of many a Botox injection is to disable these muscles.) Together, these movements create the oblique brows that so effectively heighten the appearance of vulnerability and distress, which are the signature attributes of fear. Finally, the lips of a canonically fearful face are tightened and drawn backward and slightly down. The rounded grimace that results is similar to those used by our various primate cousins to signal submission and appeasement. Vulnerability, distress, submission, appeasement—recall that these traits, which are all maximized by the movements that compose a fearful expression, are the same traits that trigger the Violence Inhibition Mechanism.

The facial muscles whose movements yield these effects are unique within the human body. Ekman has observed that the expressive muscles in the face include the only muscles in the body whose job it is to pull on *skin* rather than on bones. This is because their reason for existing is not to move the body through space for utilitarian purposes, but to contort the visible surface of the face down or up or

A fearful facial expression (here as posed by Paul Ekman) includes widened eyes, raised, oblique brows, and a grimace.
Abigail Marsh.

out in order to communicate. The resulting messages are wonderfully efficient. Not only do the contortions that yield fearful expressions work to inhibit others' aggression, but they do so very quickly. Facial muscles can contract to convey fear in a few hundred milliseconds—less time than it takes to draw in enough breath to scream. They owe this efficiency to the fact that the nerves controlling them emanate from the brain stem and midbrain, the deepest, most primitive parts of the human brain.

The rapid emergence and primitive control of human facial expressions is mirrored by the rapidity and primitiveness of how other brains respond to the sight of them. To appreciate this, let us now bestride our beam of light and tag along as it carries information about a fearful face deep into the brain of an observer.

In the nanoseconds after a person's face registers fear, much of the light striking his or her face ricochets backward from it again, radiating outward toward every other creature in the vicinity and carrying information about that face with it. That is all it means for an object to be visible: that it reflects light back toward the viewer's eyes. The regions of the face don't release reflected light evenly, though—pupils swallow it hungrily, while bright white sclera fling it nearly all back. Variations in the density and direction of reflected light enable the light to carry with it a detailed recording of all the curves and colors of the face. Let's choose one of the many available beams reflected from the sclera upon which to ride, and off we go, racing at 3 million meters per second toward the eye of a nearby human onlooker.

We reach it nearly instantaneously. After we cross the clear dome of the cornea, we pass through the pupil and into the prism of the lens, which twists us upside down and sideways as it focuses the incoming light into a sharp image. Then, after we sail through the clear jelly filling the eyeball, we make a soft, inverted landing on the fleshy retina at the back of the eye. Here we make an astonishing transformation: we are digitized. The information our light beam carries about the wide, white sclera where it originated is transformed into a digital blip of information by photoreceptor cells in the retina. Millions of these cells pulse when struck by bright white light, sending staccato messages down the optic nerve toward the brain.*

Our beam of light, now transformed into a nerve impulse, is racing away from the eye down the optic nerve on its way to the onlooker's

*As an aside, this is not the fate of every beam of light that makes it into a human eye. Some small proportion of the light that enters the eye is reflected back again, in slightly different ways by each of the various substances composing the eye. These subtleties are powerfully important. Faint gradations in the way the cornea, iris, lens, vitreous, and retina of a living eye reflect light appear to be an essential means of convincing a viewer that the being possessing that eye is not only humanlike but is actually *alive,* rather than a drawing or a humanlike doll with glassy and unconvincing eyes. It is no accident that when a person dies, we sometimes say that the light has left their eyes. The tapestry of reflected light from a living onlooker's eyes that rebounds back into the eyes of the frightened soul, then, carries with it literal glimmers of hope that another living being has borne witness to their fear.

brain. Our speed has slowed but is still blisteringly fast by human standards, about 60 meters per second. As a result, mere fractions of a second after the onlooker has taken in the wide, white sclera—plus the brows and mouth—of a fearful expression, that information has set the onlooker's brain aflame.

It's hard to overstate the effect of seeing another person's fear on a human brain. The sight changes patterns of activity in nearly every crevice of the brain, although not all at once. The first region to receive the message from the retina is a pair of evolutionarily ancient structures deep in the brain's core called the *superior colliculi*. The colliculi are two backward-facing nubs of tissue perched like pert Barbie breasts atop the brain stem. Their role is to rev up a lightning-fast response to important visual information coming in, well before the person even has any conscious awareness of what was seen. Images processed in the colliculi aren't sharp or detailed, but what they lack in precision they make up for in speed. Like a microscopic, warp-speed relay race, the colliculi pass the gist of the information carried by beams of light ("Lots of sclera! So much white!") to new fibers that extend upward to an oblong mass of neurons perched in the center of the brain, the thalamus. We zip in milliseconds to this structure, which acts like the brain's switchboard, taking in signals from dispersed areas of the brain and relaying them out again to other areas. When it receives the signal from the colliculi that the wide, white sclera of a fearful expression have been detected, the thalamus knows just where to send that information next—to the amygdala.

Findings presented in a 2016 article in *Nature Neuroscience* demonstrated for the first time that visual information about human fearful facial expressions is conveyed via this long-hypothesized ancient pathway. The researchers inserted electrodes directly into the amygdalas of eight adult humans to record activity there while the researchers showed them pictures.* They found that a mere seventy-four milliseconds after a fearful face flashed across a computer screen,

---

*This procedure was conducted in patients with epilepsy in whom depth electrodes had been placed to enable their physicians to localize the origins of their seizures.

the electrodes began buzzing with activity, signifying that the amygdala had already received information about the rough contours of the face and begun to generate a response. This is far too fast for the information to have arrived in the amygdala via any pathway other than the rapid, ancient route we have just traveled through the colliculus and thalamus. And here's the wild part: *no other facial expression* that we know about gets passed along this same privileged, speedy route to the amygdala. Not resting faces, not happy faces, not angry faces. Just fear. The mystery is: why?

Let's follow our transduced light beam for a moment more before digging into this mystery, which is itself deeply intertwined with the mystery of human altruism.

Upon exiting the thalamus, we arrive first in the amygdala's lateral nucleus, one of several semi-separate clusters of neurons within the amygdala, each of which serves a distinct role. The lateral nucleus is a sort of foyer to the amygdala where most incoming information arrives. Here we may be forced to watch helplessly as the message carried by our light breaks up, caroming in dozens of different directions simultaneously through the rest of the amygdala, then outward through the rest of the brain, as multitudes of neural cavalry are rallied to respond to what has been seen. The hubbub of activity in the amygdala following the perception of a fearful face is much greater than what follows the perception of any other expression. This is true even when the fearful expression is mostly obscured, leaving only the sclera visible. It's true even if those sclera are presented so quickly that the viewer has no conscious awareness of having seen anything at all. Dartmouth College professor Paul Whalen and his colleagues once demonstrated this by flashing just the wide, white sclera of fearful facial expressions on a plain black background to brain imaging study participants for a mere *seventeen milliseconds*—far too quickly to be consciously detected. They found that the amygdala *still* burst into a furious volley of activity—much more than when only the sclera of neutral expressions were presented. This remarkable degree of sensitivity shows that others' fear is unusually important information to the amygdala. But why?

For quite some time the thinking was that fearful expressions are important because they tell viewers that they should be fearful too. A person expressing fear is clearly afraid of *something*—a snake, a gun, the edge of a cliff. The resulting facial expression, according to this story, serves as an alarm signal telling anyone else in visual range that they may need to flee or brace for danger.

It's not an implausible explanation. Most social species use alarm signals like special calls to warn others around them of danger. Sending such calls is actually considered a form of altruism, as callers risk drawing predators' attention to themselves to warn others of danger. And just as theories of kin selection and reciprocity would predict, risky alarm calls are most likely to be used to alert family or other social group members. The auxiliary benefits of these calls extend far beyond the caller's family or even the caller's own species, though. Many species benefit from the alarm calls of even distantly related species. Birds can recognize the alarm calls of other local species of birds and even squirrels. Tropical toucanlike birds called hornbills even distinguish between and respond appropriately to the two distinct alarm calls that neighboring Diana monkeys use to warn against different types of danger (leopards versus eagles), as though they have learned the monkeys' language of fear.

Do fearful expressions in humans serve a similar purpose as these calls? Many have argued or assumed that the amygdala's robust response to fearful expressions is proof that they do. As a rule, the amygdala *does* respond rapidly to sensory events in the world that portend danger—the rippling eddy of a snake, the click of a gun being cocked, the feel of the wind along a cliff. The amygdala can learn very quickly, sometimes after a single trial, to link cues like these to incipient harm. Thereafter, when these cues are detected, cells within the amygdala fire furiously, sending urgent messages out to the rest of the brain that danger is near. My mother can thank her amygdala for the fact that she once found herself leaping into a frantic and slightly embarrassing stationary panic in front of the neighbors after a harmless garter snake slithered across our driveway. She had just recently returned home from a trip to the Amazon rainforest, where her tour group had nearly stepped on a deadly fer-de-lance lying across their path. My mother's amygdala had not forgotten how close she had come to danger. The coordinated volley of firing in the amygdala in response to danger is central to the felt experience of fear, as we know from studying patients like S.M. who lack both an amygdala and the ability to experience fear and from studying psychopaths in whom both the amygdala and the experience of fear are stunted.

So yes, it's certainly possible that amygdala responses to fearful expressions represent a learned response that these expressions signal the presence of danger. But there are also problems with this explanation. First, it's unlikely that this is the *primary* function of fearful expressions, given the impracticality of a visual alarm signal. Eyes, it seems almost too obvious to say, see only what they're looking at. What if you're looking in the wrong direction, or blinking, or sleeping, when the alarm goes off? There is a reason why fire alarms don't take the form of a little flame symbol silently lighting up in the ceiling. Ears and noses, by contrast, are always open and picking up information coming in from any direction. As a result, in most species, alarm signals take the form of barks and squeals or bursts of pheromones, not visual cues. For the same reason, the fearlike facial expressions of other primates don't really function as alarm signals. Instead, they are used to signal submission and appeasement—to inhibit others' aggression.

Amygdala responses to fearful expressions are also quite different from responses to other expressions that clearly signal threat. Angry facial expressions provide an interesting contrast. When someone is staring at you with their eyes narrowed, their brows lowered, and their teeth bared, this is clearly a threat. Anytime you see a face like this, an aggressive attack may be imminent. The face of the man who broke my nose in Las Vegas contorted exactly this way right before he hit me. But the amygdala normally doesn't respond to angry facial expressions at *all*. Angry faces actually generate even less of an amygdala response than a neutral resting face. And the amygdala's response to threatening scenes, like images of mutilated bodies, also looks different than its response to fearful faces. When the researchers who measured activity in the implanted amygdala electrodes showed their subjects images like these, they found no comparable rapid response. This almost certainly means that information about these scenes had arrived in the amygdala via a different path.

Yet another problem with the "threat response" theory is that it has difficulty explaining why damage to the amygdala impairs not just people's ability to respond appropriately to fearful expressions but their ability to even *identify* them—to come up with a name for what the expresser is feeling. When S.M. sees a fearful face, it isn't as though she knows what to call it but fails to show appropriate signs of fearful avoidance or vigilance in response. It's that she sees it and is mystified by its very meaning, like a color-blind person searching for a number in a featureless array of brown dots.

Psychopaths' seeming blindness to others' fear can also be striking. I am still haunted by a story once related at a conference by my friend and colleague, the psychopathy researcher Essi Viding, at University College London. She was testing a psychopathic inmate in an English prison and had shown him a long series of emotional faces. He was among the subset of psychopaths who are completely blind to others' fear: he got every single fearful expression wrong. Not once did he recognize the wide eyes, oblique brows, and grimace of a fearful face as signifying fear. He knew he was performing badly too. When he got to the final fearful expression in the set and yet again failed to identify it, he mused aloud, "I don't know what that expression is called. But I know that's what people look like right before I stab them."

Remarkably, this psychopath was able to recall having seen similar expressions before—and even to pinpoint the circumstances in which he'd seen them. But he was unable to discern that this particular and familiar combination of features, *even in an obviously frightening situation,* signified fear. How can this be explained? Not by the "threat response" theory.

There is another quite distinct (although not mutually incompatible) explanation for all of these findings, which is that amygdala responses to fearful expressions represent not a response to "threat" but rather a deep, atavistic form of empathy.

When a light-borne message arrives in the amygdala that the wide, distressed eyes and grimace of a fearful face have been detected, the cascade of neural firing that ensues in this structure may actually reflect a simulation of the interior state of the expresser—almost like an internal translation of the other person's fearful state. It's this simulation that allows the perceiver to understand and put a name to the expresser's state, but leaves those without functioning amygdalas drawing a blank. It's this simulation that causes faint whispers of fear to cascade down from the amygdala to a nub of brain tissue called the hypothalamus and from there outward through the rest of the body, causing most people's hearts to beat a little faster and their palms to sweat a little more in response to seeing another's fear—yet another response, not incidentally, that S.M. and psychopaths fail to show.

If all this is true—if a tiny blip of digital information carried on a beam of light can create an echo of a fear response in another person—it means that amygdala responses to fearful facial expressions represent a true conjoining of the interiors of two human brains. This would be a monumental thing. The ability to internally re-create another person's emotion, and thereby understand it, is a basic but essential form of empathy. This form of empathy is critical to the capacity to generate still more profound social responses, like *caring* that another person is frightened or distressed, and wanting to make that person feel better.

This isn't such a far-fetched possibility. A similar sort of empathic response has already been identified in various parts of the brain in response to pain. Dozens of brain imaging studies have now shown that the sight of another person in pain results in increased activity in a constellation of brain regions called the *pain matrix*. These regions include cortical regions like the mid-cingulate gyrus and anterior insula as well as deeper, subcortical regions that are also active during the personal experience of pain. The uncanny overlap in the regions that become active both when experiencing or witnessing—or even imagining—another person's pain strongly suggests an empathic response.

Even stronger support for this possibility comes from a clever brain imaging study reported in 2010 by Tania Singer, Grit Hein, Daniel Batson, and their colleagues, who examined empathic pain responses in sixteen Swiss soccer fans. All the fans were selected for being impassioned supporters of their local team. The researchers wanted to know how they would respond to the sight of pain being inflicted both on fellow fans of the local team and on fans of a rival team, using—you guessed it—electric shocks.

After each soccer fan arrived for the study, he was positioned in the MRI scanner by the researchers, who then taped customized electrodes to the back of his hand. Once the scan began, so did the shocks. The researchers measured the subjects' brain activity as electricity coursed through the electrodes and across the skin of their hands. The shocks varied in intensity, with some being very mild, and others being more painful. When the researchers analyzed the subjects' brain data afterward, they found, as expected, that activity in the anterior insula, a key component of the pain matrix, ratcheted steadily upwards as the intensity of the pain increased. The insula lies deep beneath the temples on either side of the head and is thought to encode the emotional significance of unpleasant body sensations. When those parts of the insula are active, in other words, it signals that what's happening feels bad. What the researchers wanted to know was how this same area would respond when the subject watched allies and rivals experiencing pain. Would it signal that what was happening to *other* people also felt bad?

During the study, each subject was flanked by two strangers, and all three of them were wired up to electrodes. It must have been quite a squeeze, with three grown men lined up side by side in the cramped confines of a scanner room, all of their hands positioned to be visible to the subject lying flat on his back in the scanner, peering out of it through an angled mirror. On one side of the subject sat the ally, who the subject knew was a fan of his own team. On the subject's other side sat a fan of a rival team. As the subject watched from inside the scanner, both the ally's hand and the rival's hand were also subjected to electrical shocks of varying intensities. Imagine it: You've just met a stranger, spoken to him for a few minutes, and know that he shares your love and loyalty for your favorite team. Now imagine watching his hand twitch and jerk as electric shocks jolt through it. Would you cringe with discomfort? Twitch slightly yourself? Batson and Singer's findings suggest that you might. As subjects watched the ally being shocked, activity increased in the same region of the anterior insula that was active when they experienced pain themselves, just as you would expect if the subject were simulating the ally's pain. Remarkably, though, subjects' response when they saw their rival shocked was quite different. As this stranger's hand twitched and jerked in response to the shocks, the subjects' insulas were nearly silent.

From Batson's prior research, we know that many participants in a study like this will not only experience concern for a stranger being shocked but be willing to actively help the stranger by taking on extra shocks themselves if need be. Singer and Batson again found this to be true. When given the opportunity to take on half of a stranger's remaining shocks, many participants volunteered to do so. But again, this was largely only true for their allies. When the rival received the shocks, participants were much less likely to offer to help. More, the participants' willingness to help fellow fans rose and fell in tandem with activity in their anterior insula. The more the insula responded empathically to a fellow fan's pain, the more likely it was that help would be extended.

Could amygdala responses to others' fear represent a similarly empathic response, one that could predict compassionate responses to others' distress? Perhaps. My research on psychopathy is consistent with this possibility. As I (and others) have found, adolescents and adults who are psychopathic claim not to experience strong fear themselves. This deficit seems not only to leave them callous in the face of others' fear but to impair their ability to even recognize others' fear. Other studies of large samples of adults have yielded similar findings, namely, that people who report experiencing less fear in their own lives also have more trouble recognizing it in others. It is as though a meager personal experience of fear prevents someone from even understanding what fear is, much as people who are color-blind to red and green shades cannot really understand what "red" is. The fact that psychopaths have limited personal experiences of fear *and* have difficulty even labeling others' fear strongly suggests that they are fundamentally impaired in their ability to empathize with fear—that they cannot encode and translate others' internal fearful experience into something they can understand. That psychopaths' amygdala responses to others' fear are abnormal supplies further evidence that dysfunction in this structure—a core structure necessary both for recognizing others' fear and for generating a fear response—underlies their deficits.

I should note that amygdala-based deficits impair understanding of others' fear across the board, not only when it is expressed via the face. The amygdala is essential for recognizing not only fearful facial expressions but fearful vocal expressions as well. One recent study investigating the acoustic properties of screams found that the amygdala is particularly attuned to their rough, ragged sound. Studies of patients with amygdala lesions have found these individuals to be similarly impaired in recognizing fearful vocalizations, as well as fearful body postures—even spooky music, of the kind that creates chills of fear in most people, leaves them unaffected. Quite recently, my student Elise Cardinale and I found that the amygdala is also important for identifying behaviors (threats, for example) that *cause* others fear. In a series of studies we conducted, high psychopathy scorers failed to recognize that a threatening utterance like "You better watch your back" is likely to frighten someone, and their impairments corresponded to reduced recruitment of the amygdala when considering the acceptability of uttering such a statement.

These findings are, I think, a critical piece of the puzzle. They bolster the case that amygdala deficits in psychopathy don't impair only responses to others' fear expressed via the face. If that were true, it would imply that the problem is simply perceptual and that psychopaths' problems could be solved just by giving them little strategies or clues to help them recognize others' fear—like looking for wide eyes or raised and crinkled brows. If it were only that simple! Instead, it seems, the amygdala is the final common pathway for generating a coordinated *understanding* of others' fear at a gut level, whether it is seen or heard or smelled or simply imagined. And even more importantly, the fact that psychopathic individuals struggle to understand others' fear across all these modalities provides a concrete link between empathy for fear and the experience of concern and compassion—the traits that are quintessentially absent in psychopaths.

It's not clear that the same link exists for other forms of empathy. Psychopaths are not generally impaired in understanding various other internal states, such as beliefs and goals, or even most other emotions. This reinforces the idea that empathy is not a single broad construct. There are many forms of empathy, and it is possible to possess some in abundance but lack others. Psychopaths are not really impaired in understanding anger or disgust, for example. It's not even clear whether psychopathy impairs empathy for pain. Although empathy for pain is an important social response, little evidence links a lack of empathy for pain to actual callousness. Behaviorally, there is not at this point a strong body of research showing that psychopaths have reduced experiences of pain, or that they have difficulty recognizing when others are in pain. Evidence from brain imaging studies is similarly mixed. One recent brain imaging study of psychopathic adolescents showed reduced activity in the pain matrix, but my own similar study with James Blair did not. And one study of adult psychopaths found *more* activity in the anterior insula in response to others' pain. All signs, then, point to the idea that psychopathy may be more closely intertwined with deficits in empathy for fear than empathy for pain.

Is it simply a coincidence that those individuals who are most marked by their lack of compassion and caring are *also* deficient in recognizing and responding to others' fear? Or is this actually the heart of the matter? Is the ability to generate an amygdala-based empathic response to others' distress—and fear in particular—somehow tied to the capacity for caring and compassion? If so, this might be a critical step toward understanding extraordinary altruism.

If understanding others' fear and emotional distress is essential to generating care and compassion in response to that distress, then it is clear what a peek inside the brains of extraordinary altruists should reveal. These individuals, whose attitudes toward others' welfare are so unlike those of psychopaths, whose behavior indicates that they experience unusually enhanced feelings of care and compassion for others, should show responses in the lab that are the polar opposite of what has been found in psychopaths: they should be *more* sensitive to other people's fear, and their amygdalas should be *more* responsive to fearful faces. Their amygdalas might even be larger than average as well.

Extraordinary altruists should have, in short, anti-psychopathic brains.

⁎

Another brain scan, another bunch of brain scanning subjects causing problems. This was not what I'd expected would happen on our first day of scanning altruistic kidney donors' brains to see if they were, in fact, "anti-psychopaths."

Recruiting the altruists had been astonishingly easy. Not only were they enthusiastic about participating, but they often pitched in, unasked, to help me recruit—peppering their Facebook feeds and blog posts with messages about the study and encouraging others to take part. Not one of the altruists expressed hesitation when my students explained that, for the study, they would need to travel to Georgetown for a day or two and spend over five hours completing a long battery of brain imaging and behavioral testing for a fairly paltry $150 in compensation. All of them led busy lives, and many were professionals with well-paying jobs—software engineers and bankers and physicians and marketers—but they didn't hesitate to arrange to take days off work and fly across the country to help us out. One young altruist from the Midwest told us that he was very interested in taking part but would need a few months to save up for the plane ticket. "No, no, no!" we hastened to tell him. "We'll cover all your travel costs and pay *you*—you don't have to pay anything to participate!" But think of it—he was perfectly willing to do so.

Another altruist named George Taniwaki flew in to see us from the Pacific Northwest. I know how long it takes to travel from the Seattle area to Washington, DC, as I've done it many times myself. It's a long day of travel on the best day. And this was not the best day. Sea-Tac Airport was socked in by fog and freezing rain, and his flight got canceled—twice. And then it got rescheduled twice. Many people would have given up after the first four hours of waiting. But George sat there in that airport all day, waiting, refusing to go home if there was any chance at all he might still be able to make it to Georgetown in time for his scheduled brain scan. (Most MRI scanners, including ours, are tightly scheduled, and it's nearly impossible to reschedule a ninety-minute session at the last minute.) After sitting and waiting in an uncomfortable airport seat for most of a day, and then taking a five-hour flight to DC, he finally made it to Georgetown. Then, rather than rest after his testing wrapped up the next day, he invited us all out to have dinner with him. I had never before considered whether eating dinner with a study participant following the conclusion of testing entails an ethical or scientific dilemma. Whoever heard of the question even coming up? In any case, I couldn't come up with any reason not to go, so we went, and we had a lovely evening. (I of course wouldn't let him pay for us.)

It was not this altruist, however, who caused trouble on scanning day. The culprits were the first three altruistic donors we brought into the lab, all on the same day. They were three women who had flown in from all over the country and who knew each other through the living donor community. They were excited that we brought them in at the same time so that they could spend the evening enjoying the city together. They were equally excited about the opportunity to take part in the study. One of them, Angela Cuozzo, later blogged about how fortunate she had felt to be able to participate, adding that the anticipation leading up to the weekend had been nearly "killing" her.

Perhaps all this excitement was the reason that these three forty-ish women ended up on the verge of setting off shrieking, clanging alarms all over the section of the Georgetown School of Medicine campus where the MRI suite is housed, which not even our most behaviorally disordered teenager had ever accomplished. Why? Because they were so determined not to be late for their scans. The first scan was scheduled for 9:15 a.m., the next for 10:45 a.m., and the final one for 12:15 p.m. The three were staying in a hotel about a five-minute shuttle ride from campus. So naturally, all three of them departed for campus more than an hour before the first scan. Then they got a little mixed up trying to locate the MRI suite from the spot where the shuttle dropped them off (which is completely understandable, as the suite is notoriously difficult to find). So they raced down a series of unmarked corridors in the hospital until they got so lost that they attempted to break through a secure fire exit as a shortcut. Thank goodness that plan didn't work.

They ended up backtracking and eventually managed to find their way to us, still arriving *well* ahead of time, around 8:30 a.m. They then waited patiently on the little gray couches in the waiting area, amid back issues of *Consumer Reports* and *Redbook,* until it was their turn to be tested. That meant that the last of the three scheduled subjects had arrived over *three hours early* for her brain scan. And had been worried about being late! This is a degree of conscientiousness that I feel must be nearly unprecedented in the history of psychological research. It presented me with a "problem" I had never faced before: feeling sincerely unworthy in the face of the kindness and helpfulness of my research subjects.

The scanning itself felt familiar, even if the project was anything but. Once again, my subjects watched from the dark bore of the MRI scanner as black-and-white pictures of Ekman and Friesen and all the others flashed before them one by one. Sometimes the faces wore faint smiles. Others glowered angrily. And of course, still others showed the wide eyes, oblique brows, and grimaces of fear. While they watched, the subjects held in each hand old-school black-plastic video game–looking controllers. Long wires sprouting from the controllers snaked through the MRI and across the floor of the scan room, then through holes carved into the wall of the adjoining control room, where we sat watching and where our computers recorded the subjects' responses. To keep them focused on the faces, we had instructed them to push the red button atop one controller every time they saw a man's face and to press the button on the other controller for a woman's face. Simple. Man or woman? Man or woman? Over and over and over again—more than 300 times over the course of about twenty minutes.

All the while, another roomful of nearby computers ran the scanner, manipulating the massive magnetic field surrounding the subjects' heads. *CLACK-CLACK-CLACK-CLACK-deedeedeedeedee-deedeedee,* rumbled the scanner, its machinations causing tiny charged particles inside the altruists' brains to tumble and spin on their axes. Inside the scanner, the birdcage-shaped coil encircling their heads collected the faint radio signals created by all this tumbling and spinning. I could picture the pink, two-centimeter ovals of the amygdalas pulsing deep within their brains, waiting for their cue. The subjects hadn't been told to pay attention to the expressions on the flickering faces, but no matter. Flash! A bright, white sclera leapt from the screen. Did the amygdala blaze to life a few dozen milliseconds later, its cells pulsing out a Morse code message to the rest of the brain? *Look sharp! Someone's scared!* If so, it would momentarily increase this structure's fuel consumption by a tiny amount, no more than 1 percent or so. But that would change the way the protons within it danced and spun just enough for us to measure—enough to inform us how an extraordinary altruist responds to others' fear.

As friendly and helpful and generous as the altruists were, and as patiently as they lay in the scanner pushing the red buttons, many of them didn't hesitate to tell me—tactfully—how wrongheaded the study was. Harold Mintz was perhaps the most vocal of them. Harold's story is an unusual one, even in the world of altruistic kidney donors, because, like Sunyana Graef, he had never heard of anyone donating a kidney to a stranger before he came up with the idea on his own. He first had the idea around the time Graef did, in 1998 (what was in the air that year?), at which time he was living in Arlington, Virginia. Unlike Graef, he didn't live near a transplant center willing to remove his kidney and give it to a stranger. He tried contacting the National Kidney Foundation, but all he got in return was a stack of pamphlets in the mail that explained how to donate his organs after he died. *Hey,* he thought, *I'm not dead yet.* So he called them back. He tried to clarify: "I'm just curious about donating *now* to somebody here in the DC area."

A long silence followed. Then:

"You can't do that. It's illegal."

But they took down his name and number and said they would call him back if anything changed.

Yeah, sure, Harold thought.

But something did change. That same year, the Washington Regional Transplant Consortium was in the early stages of developing what would become the first community-based living organ donor registry. They got Harold's contact information from the National Kidney Foundation and called him back two years later to say that they were launching the program. Did he still want to donate? He jumped at the chance. After a long bout of psychiatric and medical screening, Harold became the very first person the program approved to donate.

Knowing Harold, this doesn't surprise me at all. I find it hard to imagine meeting with him and denying him the chance to give away his kidney. He looks like a cigarette billboard cowboy, complete with a wild shock of graying hair and a fulsome grizzled mustache, and he has the charisma of a preacher. He's one of those people who seems to make the air around him sparkle. I've shown clips of my interviews with him at conferences, and attendees will circle back to me years later to ask about the "mustache guy."

Harold describes his decision to give away a kidney as if it's the most obvious, the most straightforward decision a person could make. When asked to explain why he donated, he responds with a question that most people find easy to answer, which is, "Would you give a kidney to your mother to save her life?"

Essentially everyone answers yes.

He'll nod, then scribble down a few letters on a piece of paper. He'll then ask, "Okay, why? *Why* would you donate to your mom?"

Everyone, he says, replies exactly the same way. I did as well when he asked me: "Because she's my mom."

Harold then flips the paper around to show the letters "BSMM" written on it: Because She's My Mom. He already knew what they would say.

And what people say is telling. The response "because she's my mom" is not really an explanation at all. It's hardly better than just saying, "Because." It's not a description of costs and benefits, or an exegesis on the details of what the surgery might entail and the likely long-term outcomes of kidney disease. The answer is much more primitive: she's my mom, she's going to die otherwise, so I'll give her my kidney and worry about the details later. In this way, in this response, we are almost all the same.

Then Harold will push further: "So we got that, you'd do it for your mom. Okay, how about your sister or your brother? How about your best friend, who's not related? How about your teacher, or your boss?" He extends the circle further and further out, challenging you to tell him when you would stop caring enough about another person's life to give them your kidney. "What if," he asks, "that person is going to die *next week* and you're the only person who can save them? There is somebody dying *right now* while we're having this conversation, where the doctors know exactly what to do to fix them. We can actually stop somebody from suffering—from losing what they might have."

These questions echo a conversation Harold had while being screened for his donation. He had asked the transplant team, "If I don't give my kidney to somebody this week, will someone die waiting for it?"

"Yes," they told him.

That sealed the deal. For him, Harold says, taking all of this into account, donating his kidney really wasn't a choice. It was an opportunity. "Because someone is going to die" was for him just as obvious and simple an explanation for donating his kidney to a stranger as "because she's my mom" is for everyone else.

On December 12, 2000, in an operating suite at Georgetown University, steps away from where we would later scan his brain, a team of surgeons removed Harold's left kidney and stitched it into the abdomen of a young woman named Gennet Belay, a wife, mother, and Ethiopian immigrant who at the time had only 6 percent of her kidney function remaining. In a short film called *1-800-Give-Us-Your-Kidney,* Belay's husband recalled that her doctor had told her at one point she had only three days left to live. Harold's kidney started working to filter her blood almost immediately after the transplant and continues to do so to this day. Harold sends it a birthday card every year. Belay considers the day of the transplant her own "re-birthday"—the day Harold returned her to a life free of disease and medical complications.

Harold recalls feeling no anxiety or fear or doubt leading up to the donation, only excitement and a little frustration that the whole process took so long to get under way. He never once reconsidered his decision. And without question, he would do it again if he had the opportunity, which of course he won't have—even the most impassioned altruist has only one spare kidney to give.

I once asked Harold, as I ask all the altruists we meet, to tell me why he thinks he is one of the less than 0.001 percent of the population for whom giving a kidney to a stranger is just as obvious as donating to their mother would be for most other people. What is different about him? His response is vehement, and it echoes the words of Cory Booker and Lenny Skutnik. "I'm *not* different. I'm not unique," he maintains. "Your study here is going to find out that I'm just the same as you."

As he sees it, our study wasn't even asking the right question. Altruistic kidney donors like him are, in his view, ordinary people who are in the right circumstances at the right time, with the right information. Being portrayed as any kind of hero frustrates him to no end. He has told me unambiguously and repeatedly that he is not.

Perhaps he's right. As a scientist, I try to keep my mind open to any possibility not conclusively ruled out by the evidence. It's possible that, as Harold maintains, donating a kidney is mostly circumstantial, that nearly anybody with the right combination of knowledge and prior personal experiences would be similarly motivated. Many of the kidney donors we've worked with believe this is true. The first three altruists we tested—the ones who tried to break through the emergency exit door—all gave responses to this effect when I asked them, in separate interviews, why more people don't donate:

"I would say: information."

"Lack of education."

"It's just not knowing."

If more people knew about donation, they averred, more people would donate.

Of course, this must be true to some extent. Every altruistic kidney donor was once *not* an altruistic kidney donor, but rather someone who, in most cases, had never even heard about nondirected donations. Then, eventually, each of them discovered that it was possible to donate a kidney to a stranger, or they learned how many strangers out there needed a kidney and this new information was the precipitating event that led them to donate. What distinguished their predonation and postdonation selves was the information they possessed. That altruists explain their donations this way is quite consistent with a general phenomenon known in social psychology as the *actor-observer effect:* people tend to explain others' behaviors with reference to internal causes like personality, but explain their *own* behavior with reference to external causes—in this case, the acquisition of new information.

But that seems unlikely to be the only reason that people give strangers their kidneys. For one, the rest of us respond quite differently to exactly the same information. If you're like me, you might have followed along with Harold's line of questioning for a while. You'd donate to your mother for sure. So would I. Your brother, yes. Best friend, okay. But somewhere along the way there's a shift. The answer no longer seems so obvious. Your neighbor, or your teacher? Your boss? *Maybe.* For me, these decisions start to feel different somehow, less instinctive. The details I'm happy to shove aside until later for my mom's sake rapidly return to the forefront when I think about donating to someone more distant. By the time I get all the way to thinking about donating to someone I've never met . . . there's just a blank. Nothing about that decision feels obvious at all.

This is a common response even in people who are deeply and painfully familiar with the desperate need for donor kidneys and the lifesaving power of altruistic donations. In the film about Belay's transplant, her husband mused, "As far as Harold is concerned, it's not easy risking your life to save somebody else who you don't even know, who you have never seen. I was asking myself, 'Could I do that?' And my answer was—no. I know what it means. When you have only one kidney, you are risking your life. He must have . . . a special heart."

A special heart—or a special brain?

It took more than a year, but we eventually collected enough data to find out. In all, we scanned the brains of nineteen altruistic kidney donors while they viewed the various emotional facial expressions. They included Harold, Angela Cuozzo and the other women who were so determined not to be late, George Taniwaki, who had flown in from Seattle, a real estate consultant, a mechanic, and a dozen others. In addition to measuring activity in their amygdalas and elsewhere during the scans, we also collected anatomical scans that would inform us about the size and shape of all the various structures within each of their brains, including their amygdalas.

For comparison, we also collected identical data from twenty control participants of the same average age, years of education, IQ, and other variables as the altruists. The only other requirement for controls was to have never donated an organ to anyone. That's most of the population, of course. A dragnet of any single city block within a mile of Georgetown University would probably yield twenty such adults, so you'd think it would have been a cinch to find them and test them. You'd be wrong. It really reinforced for me how incredibly abnormal (in a good way) the altruists' eagerness to participate was—especially given all the travel hassles involved—that it took us *twice* as long to find enough controls for the study as it did to find the same number of altruists. This is despite the fact that the controls were all recruited locally and didn't have to travel anywhere, and despite the fact that there are literally *100,000 times* as many of them as there are altruistic kidney donors. (I remain, of course, fantastically grateful to all of the controls who did participate—the study would have been impossible without them.)

My students analyzed our data using the same painstaking process as usual. Hours and hours and hours of computers whirring away to turn gigabytes' worth of raw binary code into three-dimensional images of human brains that flickered and glowed with activity. Our final analysis aimed to see how much more active the amygdala was when the two groups of subjects looked at fearful expressions as compared to neutral expressions. The moment of truth arrived again. When we compared the altruists' brains with those of controls—who were like the altruists in every way we could think to measure except for not having donated a kidney—what would we find?

Bingo. There it was again. Glowing like a little star. Half a cubic centimeter or so of flesh inside the altruists' right amygdala had recruited more blood to fuel its activity after fearful expressions appeared.

Now, all we really knew from this finding was that cells somewhere in the altruists' amygdalas—was it the lateral nucleus? some other nucleus? we couldn't say—were more active when they gazed at a stranger's fearful expression than when they gazed at a neutral expression. Was this the legitimately empathic response we suspected? Or was it something else, like a response to threat?

One clue came from the results of a comparable analysis we conducted to evaluate altruists' and controls' responses to angry expressions. This time we found that the pattern was reversed—the amygdala was *less* active in altruists than controls when they saw angry faces. This pattern isn't consistent with the idea that altruists' amygdalas are simply "threat detectors" when it comes to expressive faces. A useful contrast can be drawn with people who have clinical anxiety disorders like generalized anxiety disorder or generalized social phobia. When these people view facial expressions in the scanner, they show heightened amygdala responses to a whole range of negative stimuli, including fearful expressions, angry expressions, and other expressions like contempt. Anxious people are overly vigilant to possible threats and danger, and their amygdalas tend to be perennially hyperactive; in them, amygdala activity to a whole range of cues may well reflect some form of threat detection. But the fact that the altruists were more sensitive *only* to fear suggests something else was afoot.

Another clue as to what that "something else" might be came from additional data we had collected after we wrapped up brain scanning with each subject. After they'd had a little break and some lunch, we invited them to come back to our lab for questionnaires and some computer tasks. One of these was an emotional face recognition task that contained both angry and fearful facial expressions. When we analyzed how well the two groups of subjects identified the emotion conveyed by these two expressions, our findings mapped neatly onto our brain imaging data and echoed the emotion recognition data I had collected for my dissertation a decade before. Compared to controls, altruists recognized fearful expressions relatively better. By contrast, they recognized angry expressions relatively *worse*. It was only their empathic accuracy for others' fear that was better than average. This finding reinforces the idea that empathy can take many forms and that each form is driven by partially distinct processes, such that it is perfectly possible to have a high degree of empathy for others' fear but not their anger. When it came to the altruists' empathic sensitivity for fear, the signs pointed to the amygdala driving this accuracy. Across our subjects, we found a strong correlation between how active a subject's amygdala was to fearful expressions and how well that person recognized these expressions later on.

So far, our altruists looked remarkably like "anti-psychopaths": they recognized others' fear relatively better than controls did, and this ability seemed to correspond to a more robust response in the right amygdala to these expressions. What about the final feature of psychopathy we had considered—the overall size of the amygdala? My student Paul Robinson crunched the numbers to generate mockups of the average shape and size of all of the subjects' amygdalas, although I was a little dubious about whether this effect would pan out, to be honest. But the results clearly showed that in this respect as well, the altruists appeared to be the opposite of psychopaths. Their right amygdalas were physically larger than those of controls, by about 8 percent. The significance of this effect held up even after controlling for something we had not predicted, which was that the altruists' brains were larger overall than controls' brains.

Despite what Harold and many of our other subjects believed, something about the altruists' brains really was special. Altruists, it seems, may be more strongly affected by the "field of force" that promotes compassion because the sight of someone suffering affects them more strongly than it affects the average person. They appear to be equipped with just a little more of the three features that psychopathy research has identified as being essential to ordinary compassion—the basic neural hardware required to be sensitive to signs of extreme distress in others, which highly psychopathic people lack. And that little something extra—the extra sensitivity, the extra activity, and the extra volume—may provide altruists enough of a boost to move them past ordinary levels of compassion into something extraordinary.

Perhaps these small changes can help to explain why the strange blank feeling I get when I contemplate donating a kidney to a stranger feels like *something* to an altruist. As the altruists tell it, that same sense of certainty and purpose I feel when contemplating donating a kidney to someone I love is what most of them felt when they first contemplated donating to someone they'd never even met, that sense of, "Okay, you can have my organs . . . it's a no-brainer," as one young kidney donor from Arizona put it. Or in the words of another altruist, "I had no particular reason other than, like I said, you see someone drowning, you are going to pull them out of the water. . . . I knew you gotta help when someone suffers." Simple as that. Clear as Lenny Skutnik diving into the Potomac. Instinctive as Cory Booker racing into a neighbor's burning house. Fast as my roadside rescuer hitting the brakes. The fact that their choices seem to boil down to gut-level, intuitive feelings all made so much sense once we discovered differences in the altruists' amygdalas.

The amygdala, as brain structures go, is pretty deep under the hood. It can respond to stimuli you have no conscious awareness of—bright white sclera flickering for a few milliseconds; the smell of someone's sweat when they're frightened—and change your behavior and ongoing thoughts very quickly in response. If something very fast and very unusual is happening in this ancient structure when extraordinary altruists witness or contemplate someone else's distress, is it any wonder they have trouble articulating what it is? It may legitimately feel like the decision is a "no-brainer" when there is no easy access to the part of the brain where these critical calculations are taking place, although perhaps a better term for it would be "deep-brainer."

The fact that altruists show heightened empathic responsiveness to others' fear also reveals an important truth: there is a critical distinction between being fearless and being brave. Many psychopaths are genuinely fearless, and as a result they have difficulty understanding others' fear. That altruists are so empathically responsive to others' fear suggests that, rather than being fearless, they are unusually sensitive to fear. Recall the half-dozen different ways in which Cory Booker described the terror he felt while rescuing his neighbor from a fire. And Lenny Skutnik, who saved a stranger from an ice-choked river full of wreckage and jet fuel, later found himself overwhelmed by nerves during an interview with Ted Koppel. I have asked dozens of altruistic kidney donors if they consider themselves to be fearless or low-anxiety people, and the answer is nearly always an emphatic "no." Almost none of them engage in classically risky activities like skydiving, which, you will recall, is actually *less* risky than donating a kidney. Sunyana Graef said, when I asked her about risk-taking, that she had "gone parasailing once"; otherwise, she definitely didn't engage in any risky behaviors, she said, because doing so "wouldn't be right." More than one of the altruists we've studied was afraid to fly, judging by their requests to take anti-anxiety medications during the flight to Washington. (Unfortunately, we had to request that they not do so, as the drugs' residual effects could have interfered with the brain scans.) During her interview, one altruistic donor from New York reeled off a list of small, everyday things that she worried about, from being late with her rent check to running out of gas on the freeway. And another from San Francisco said that for most of her life she had been afraid of "everything in life . . . absolutely everything."

Her words reminded me of those of the heroic Civil War battlefield nurse Clara Barton, who reflected in her autobiography: "Writers of sketches, in a friendly desire to compliment me, have been wont to dwell upon my courage, representing me as personally devoid of fear, not even knowing the feeling. However correct that may have become, it is evident I was not constructed that way, as in the earlier years of my life I remember nothing but fear."

Our findings suggest that Barton's words reflect a deep truth, which is that true selfless heroism emerges not from the absence of fear, but because of it. People who rescue strangers from fires or drowning or who donate their kidneys to strangers seem to be acutely aware of what it means to be afraid. And this awareness may be in part what moves them to help others. Their bravery lies in their ability to recognize and empathize with acute distress, while simultaneously overcoming or overriding their own fear in the face of danger. They are able to respond altruistically because, even while they empathize with others' fear, they do not allow fear to flood their own system and prevent them from acting to help.

How on earth do they pull this off? At least on the face of it, it doesn't appear that they make any conscious efforts to suppress their own fear; indeed, altruistic kidney donors often report being surprised to discover how they were feeling as their donation date approached. When I have asked donors what their dominant emotions were right before they went under anesthesia, the response I get the most often is Harold's answer: "Excitement." Another young altruist in his twenties said that right before his donation, "I was really excited for it actually. I do not know why. I do not know what was the exciting aspect about it. I think just knowing that I was gonna be able to help somebody so much was really cool to me and everyone was so worried, like I was going to die on that table. Everyone was like, 'Why you are doing this? You are going to die!' And for me, I almost felt strange not being worried about it."

Many donors have even said that they felt an unexpected sense of peace or certainty. One altruist explained, "I would not consider myself fearless. I do not take a lot of risks. Somehow, I never thought this was a risk. I just knew from the beginning that I was going to get through it fine. I do not know why I knew that, but I knew that." Sentiments like these echo the words of Lenny Skutnik, who, again, was not a generally fearless person, but recalled feeling no fear before he leapt into the Potomac—an extraordinarily risky and painful choice—but only a calm conviction that everything was "going to be all right."

How is it that people who have normal—or even higher than normal—sensitivity to fear and anxiety find themselves feeling anywhere from calm to excited before voluntarily undergoing significant pain and risk to save another person's life? What neurobiological process could conceivably transform an act that is objectively risky or costly into one that generates feelings of calm, even positive excitement? The answer to this question may be the final piece required to understand the puzzle of extraordinary altruism, and it may boil down to the basic mechanisms that underlie the capacity for care.

6

THE MILK OF HUMAN KINDNESS

HER WIDE, PALE belly scrapes across the sand as she heaves herself forward. Her thin, formless legs struggle for purchase, but her shoulders are strong. Again and again she pulls herself onward, gaining a few inches each time. The going is slow. The sand slopes unrelentingly upward, and her body weighs over 200 pounds. It leaves a ragged trench in the sand behind her, one that is easily visible even by moonlight. She stops often to rest. At last, about half an hour after first emerging from the ocean, she reaches her destination: the edge of the north Florida beach where the flat-packed sand yields to sugary heaps shaded by stalks of sea oats and creeping purslane. She has seen this spot before, although not for many years, and never from quite this vantage point. But she's certain it's the right place. After a pause to recoup her energy, she settles herself in and begins to dig.

She isn't seeking treasure, nor would she know what to do with it if she found any. She's here to build a nest. Inside, she will deposit dozens of rubbery, Ping-Pong ball–shaped eggs containing the embryos of her children, just a few of the thousands of loggerhead sea turtles she will produce over her lifetime.

The nest ready, she positions herself above it. The leathery tube of her cloaca bulges as the first egg of the clutch pops from it, shiny with fluid, bouncing off the side of the pit before coming to rest at the bottom. Dozens of siblings tumble out after it in succession, soon burying it. She never gives the gleaming pile so much as a glance. When her oviduct is empty and the pit is full, she flips sand back over the nest in a chaotic spray that coats her own head and shell as well, until the eggs are hidden from view and protected from the sun and wind and gulls and crabs. Afterward, she tamps the pile into a firm lump with the underside of her shell. It doesn't look like much, but all this spraying and tamping is a significant moment in the lives of her offspring, as it represents the only care they will receive from their mother. Her job complete, she drags herself laboriously back down to the sea, where she disappears beneath the surf along with the last thoughts she will ever give to her babies.

Assuming this is a typical loggerhead nest, it holds about 115 eggs. Many of them will never hatch, some because they were never fertilized. Others will succumb to any of the various hazards that can befall a two-inch-long baby developing unattended in a shallow pit in the sand for several weeks, including being killed by fire ants or crabs or being discovered by poachers or overheating or getting flooded during a storm surge. Of the ones that do hatch and manage to dig their way straight up through a foot of eggs, fellow hatchlings, and packed sand, plenty will never make it to the water. Some will become disoriented by the bright lights of beachfront hotels and die of dehydration. Others will get picked off by shoreline predators like raccoons and gulls. Those that do reach the ocean will have a fighting chance at survival, but they remain so tiny and so vulnerable and face so many obstacles that nearly all of them will ultimately die young too. It's estimated that only one of every 1,000 loggerhead babies reaches adulthood—that's a one-in-ten chance that even *a single hatchling* from this nest will survive to maturity. No wonder their mother didn't bother getting sentimental about them.

I encountered these hatchlings several weeks later and was not as lucky. If anything, "sentimental" underplays my response to them.

This fact—that the same baby turtles could be forgotten and left quite literally for dead by their own mother and also be the object of profound and faintly embarrassing sentimentality from a passing human—is deeply intertwined with, and essential for understanding, the origins of humans' capacity for altruism.

It was early July when my family and I recently visited my in-laws in Ponte Vedra Beach, Florida. That happens to be hatching season in a prime loggerhead nesting area. Turtles who lay eggs there are fortunate to have all of their nests recorded and monitored by the stalwart volunteers of the Mickler's Landing Sea Turtle Patrol, a local conservation group. Every spring and summer, volunteers patrol four miles of coastline for signs of turtle nesting. When they spot telltale crawl marks ending in packed-down lumps of sand, they pound wooden stakes into the sand around the nest and scrawl on them the estimated date the eggs will hatch. The patrol's policy is to wait until three days after the first hatchlings emerge from a nest, then dig up anything that remains to log how many eggs were laid and the fate of each. Was it fertilized? Did it hatch? Did the turtle successfully exit the nest? The best part of a dig is that anywhere from a few to a few dozen turtle hatchlings that have not yet dug their way out are usually unearthed, and are then released by the patrol to make their way to the sea.

My mother-in-law Krista brought me to watch this process with our young daughters on a sweltering Saturday night when a loggerhead nest was scheduled to be dug up. It was about the saddest thing I've ever witnessed. Two patrol members found the appointed spot in the sand, and we all circled around. A few hatchlings had emerged three days prior, so hopes were high. One volunteer carried a small red pail into which any new hatchlings would be placed for monitoring before being released. The other volunteer began to dig. Her gloved hands gently carved through the sand as she scraped away layer after thin layer, like an archaeologist, or a sculptor. When the first glimpse of smooth, pale shell appeared, we all held our breath. But it was a dud—an egg that had never developed. Its saggy profile and yellowish tint gave it away. After a few more duds, the first sign of hope appeared. The fragment of a clean, white shell was a sign that at least one hatchling had emerged alive. More fragments followed, as well as some fertilized but unhatched eggs, which the volunteer placed to the side. Eventually, the volunteer's sliding hands revealed the thing we had most hoped to see: a thin sliver of black no longer than the first joint of my finger—the flipper of a baby loggerhead. But the flipper did not flip. It lay motionless in the sand, lifeless as the shell fragments surrounding it. The volunteer covered it with her hand, hiding it from the ring of hopeful children gathered around, then pried the rest of the body from the sand with her fingers. As she lifted it up, the flipper swung out from under her curled thumb. Carefully, she moved the body to a trough carved in the sand for counting purposes. Another trough already held the eggs, duds, and fragments that had been uncovered. This trough would hold only death.

After a while, there was no point trying to hide the bodies anymore. One still, dark form after another emerged from the sand, turtle upon turtle in various stages of decay—perhaps ten in all. A few were putrid and hardly recognizable. But several, like the first, were still perfect in every feature. My six-year-old daughter, to her credit, was not frightened by them, only curious, so we moved around to the trough to get a better look. I stared for a long time at the first one to be dug up. Its tiny chin rested peacefully on the sand beneath its beaky nose and the large dark whorls of its closed eyes. Its shell was a mosaic of irregular black pentagons fit neatly together. Its front flippers were outsized, like the long limbs of a foal, and their trailing edges bore intricate notches and ridges that I imagine would somehow improve hydrodynamics. Such a perfectly formed thing it was in every detail; a sculptor would weep with joy to create an object of such beauty.

But of course a mere sculpture's marvels are all external. What lay inside this turtle was, if anything, even more wonderful. Within its chest blossomed an intricate four-chambered heart and lobed lungs, life-giving machines beyond the reach of anything science can yet replicate. Behind its closed lids lay eyes of such complexity that

Darwin himself wrote that it seemed "absurd in the highest possible degree" to imagine such a thing evolving through natural processes. A wholly formed brain had pulsed inside its skull—millions of neurons linked together in intricate webs ready to support this turtle's ability to escape its shell and dig and breathe and learn—maybe even feel. All built for nothing. All built so that the sum total of all of these wondrous parts could lay moldering under the setting sun, a checkmark in the patrol's "dead hatchlings" column. The "live hatchlings" column stayed empty. The horrible waste of it all was overwhelming.

We returned the next evening with hopes for something less depressing, but the first of the two nests to be dug up was even worse. Whereas the first nest had at least born signs of hope and life, including intact fertilized eggs and many more shell fragments than dead hatchlings, the second was an awful avalanche of rotting hatchling carcasses, so many so badly decayed that it seemed unlikely that more than one or two had possibly made it out alive. The verdict was that fire ants from a nearby colony had attacked the nest and killed all the eggs and hatchlings, some still only partway out of their shells.

"This," said the stoic volunteer by the time she had finally dug to the bottom of the whole horrid mess, "is not a happy nest. I'm so sorry."

One more nest remained to be dug up before we were to return home to Washington, DC. I hardly had the heart to watch the whole thing again, but the girls begged to stay, so we did. I watched stonily, trying to detach myself from the process this time. The first scrap to appear in this last nest was positive: a shell fragment. More followed. No duds yet, no dead hatchlings. "Don't get your hopes up, don't get your hopes up," I muttered to myself sternly. Then—unmistakable movement! The crescent of a tiny dark flipper flung itself from the sand and scrabbled the air. A thumb-sized head followed, craning toward its first glimpse of the sky, its round, black eyes blinking back the light and crumbs of sand. Jubilation ensued. Children shrieked. My daughter and sister-in-law laughed and hopped up and down. I wanted to laugh myself. It was as winsome and intricate as all the poor motionless ones we'd already seen, but to see its limbs flapping with energy and life was pure joy. More soon followed—eight live hatchlings in all. That is a smallish number, but it felt like a cornucopia. The crowd welcomed each turtle into its sandy new world with a cheer. The volunteers placed them one by one into the red bucket, where they stretched their wrinkled necks toward the setting sun and stepped on each other's faces while the patrol volunteers finished digging out the nest and making final tallies. They were impossibly cute.

Then it was time to release them. The volunteers aim to interfere minimally with the natural process, so they wouldn't bring them to the ocean, a thirty-second walk away at the most. Instead, they gently tipped the bucket up next to the nest. The hatchlings had to do the rest themselves.

Oh, the agony and delight of watching them struggle to reach the water! They paddled over the sand like windup toys, their flippers moving in perfect metronomic rhythm—flip-flip, flip-flip, flip-flip, flip-flip. They slowed only slightly as they moved over and around the flotsam and hillocks they encountered. Their certainty of purpose was comically at odds with their miniature bodies. They fanned out as they moved toward the ocean, and the volunteers struggled to keep an eye on all of them and the phalanx of eager watchers ushering them onward, many of us holding our arms out at protective angles. One poor hatchling lost his way repeatedly, heading each time for a thicket of children's legs. More than once he narrowly avoided being crushed by an errant flip-flop. Each near-miss yielded groans of panic from the onlookers and shouts of: "Move back! Move back!" It was so hard to keep track of all eight. Any attempt to avoid stepping on one risked crushing another. My stomach churned at the thought of one being trodden on—or worse yet, treading on one myself by accident. I didn't think I could stand seeing another of their tiny bodies go limp, not after having watched them struggle so mightily to reach the lip of the vast churning ocean that was their only hope of survival.

Despite our efforts, tragedy befell the hatchling that kept getting lost. He moved too far from the group, and a gull seized its opportunity to snatch him up. He was carried high in the air while a few people who'd seen it happen screamed invectives at the bird. Was it our screaming? Did the turtle put up a fight? For whatever reason, the gull dropped him. He plummeted down to the sand, fifteen feet at least. A collective gasp. Then, "He's still alive!" shouted a triumphant volunteer. Not only alive but undeterred, he righted himself and returned to his mission. The onlookers closed ranks around him, ushering him the last few yards of his journey. He was the last one in, but he too made it to the ocean at last. When he reached the waterline and the first gush of bubbling froth hit him full in the face, I swear I saw a look of astonishment in his eyes. He froze for a moment at the novelty of it; yet another incredible event overwhelming his newly formed senses on this, his first day in the world. He quickly recovered and started to paddle, and soon the rushing tide had carried his small form from view.

Walking back to the car afterward, the four of us felt foolishly proud of the hatchlings for making it to the ocean, for having already beaten so many of the odds against them. Everything about this feeling was ridiculous, of course. They weren't ours. They were reptiles who had been buried in the sand by a turtle, were monitored and dug up by patrol volunteers, and would have made it to the ocean on their own whether or not we had been personally cheering them on and yelling at people not to step on them. If I had happened to visit Florida on a different weekend, I would have never known they existed. So why had I—and thirty other beachgoers—become so instantly protective of and invested in them? So willing to spend time and energy taking what feeble measures we could to ensure that they made it to the water alive—shouting at each other and even at a sea gull (who was just hungry; why did we take the turtle's side?), holding our arms out like beach bouncers as though this would somehow keep the turtles safe? Why did I feel so delighted watching the hatchlings' paddling flippers and their sweet, beaky faces, and so terribly hopeful that they might continue to beat the odds and survive? Why did I feel so—there is no other way to put it—*sentimental* about them when their own mother spared not a single thought for any of them after she had tamped her final tamp over their nest and most likely would have ignored them if she had happened to be on the beach again that evening and seen them scramble past?

The answer traces back many millions of years. The short version is that I am a descendant of creatures called cynodonts, and loggerhead turtles are not. The longer version of this answer may provide the remaining pieces of the puzzle that is extraordinary altruism.

Loggerhead turtles are an ancient species, and a successful one in the scheme of things. They have been leaving their eggs under lumps of sand on beaches around the world for some 40 million years. They and six other existing species of sea turtles, collectively known as *chelonioidea,* can trace their ancestry back to the progenitors of all modern turtles who emerged more than 200 million years ago during the Triassic Period. That means turtles predate even dinosaurs and birds, who are the dinosaurs' last surviving descendants. (I know the idea that birds and not turtles are the true heirs of the dinosaurs can seem a little silly—especially if the birds you're thinking of are pigeons or parakeets—but just watch a great blue heron stalking fish in a stream sometime, its scaly legs and clawed feet parting the reeds while its fierce, pointed head swivels atop its long neck, and it will seem much more obvious.)

Humans and other mammals are the descendants of neither dinosaurs nor turtles, but of hamsterlike creatures called cynodonts whose lineage diverged from other four-limbed animals about 250 million years ago. Like modern mammals, cynodonts were furry and warm-blooded, but they also laid eggs. Together, these traits left them backed into something of a corner when it came to reproducing. To produce warm-blooded babies able to maintain their own body temperatures after hatching would require that the babies be very large—so large that gestating and excreting eggs big enough to contain them would have killed their tiny mothers. The only other option was to produce very small eggs, out of which very small and developmentally immature babies would hatch. But such babies would be *so* immature that they would be incapable of even supporting their own metabolism without someone keeping them constantly warm and nourished. They would be, in other words, *altricial,* which is the biological term for babies born immature, helpless, and dependent. That this word sounds like "altruism" is not a coincidence. Both words are derived from the Latin *alere,* meaning "to nourish." Altricial babies can be contrasted with those who are *precocial,* or developmentally mature and self-sufficient. Human babies are quintessentially altricial, as are the babies of many other mammals and nearly all birds, whereas sea turtles and other reptiles and fish and any other animals capable of showing anything like "certainty of purpose" within minutes of being born are precocial.

So this was a pickle. How could cynodont mothers simultaneously keep their newly hatched babies both warm *and* fed around the clock? There are only two viable solutions to this problem, and cynodonts came up with one of them. Everything that we know about modern and ancient mammals points to the conclusion that the critical evolutionary development that allowed cynodonts to flourish was that cynodont mothers developed the ability to keep their tiny, altricial young both warm and fed by *turning their bodies into food.*

In other words, they made milk.

Milk is among the more spectacular evolutionary developments ever to come along. It's easy to take for granted that mammalian mothers literally dissolve their own flesh and bones and excrete it as food through their nipples because we don't know any differently. But imagine if you one day discovered that you could shoot hamburgers out of your armpits at will. That's basically how incredible lactation is. Except milk is even better nourishment than a hamburger. It contains water and salts essential to sustaining life; nutritive fats, proteins, and sugars; various other vital substances, from calcium and phosphorus to the indigestible oligosaccharides that serve as prebiotics in the gut—all served perennially fresh, liquefied, and prewarmed directly from mom.

This elixir is directly responsible for the emergence of all the diverse modern-day mammals that populate the earth today, humans included. We could never have survived without it. It's little wonder that mammals were named for the very glands that produce it. Milk allows metabolically greedy mammalian babies to be born small and underdeveloped and to survive their vulnerable infancy by receiving all of their nourishment from their mother while staying warm and expending almost no effort—not even chewing. It allows mammalian babies to thrive in nearly any ecological niche, since they need neither to compete with adults for food nor to find an alternative—and likely inferior—food source. This makes milk a likely reason that so many mammals survived the aftermath of the cataclysmic asteroid strike that killed off the dinosaurs. Milk also allows mammals to sustain unusual and beneficial growth patterns, like disproportionately rapid growth of the head and brain soon after birth. And finally, milk ultimately transformed the social life of the mammals who consume and produce it as well. Consuming milk is no mere luxury for mammalian babies but a necessity, so milk is the reason that mammalian babies remain dependent on and attached to their mothers for weeks or months or years—and the reason that their mothers remain attached to their offspring as well.

Milk is, in other words, the prime mover behind many of the psychological, behavioral, and social features that separate mammals from all the creatures who came before them, and from many who came after as well. This was an inevitable outcome, since the ability to produce milk would be a useless adaptation in isolation. It's only worthwhile if accompanied by a suite of other changes that ensure that infants actually benefit from the milk.

What is the first and most essential of these changes?

Love.

Or, if it makes you uncomfortable to think of the same animals that we eat for dinner and experiment on in laboratories experiencing love, call it caring, which is love's behavioral expression. It doesn't really matter what you call it—it all boils down to the same thing.

And that *thing* is that producing milk to sustain altricial offspring is useful only if the mother stays in frequent and very close contact with her offspring to feed them the milk—literally pressing her body against them or cradling them or hovering directly above them for stretches of time every day for weeks or months to allow them to drink their fill. And no mother is going to do this every day, without fail, for every infant or litter of infants she produces, breeding season after breeding season, unless some powerful motivational tether keeps pulling her back to her babies again and again when she could be out doing something easier or more fun or interesting than hanging around being drained of her own liquefied flesh and bone. And that tether is love.

That love underlies the urge to be in physical contact with the newborn offspring as much as possible. To smell their intoxicating smell, to gaze at the endless wrinkles and curves of their small, strange bodies. To find their simple presence wildly reinforcing, and to feel the fierce longing of an addict when separated from them. To want to stroke them or cradle them or lick them or nuzzle them (depending on the species and available appendages) and protect them from the amorphous terrible dangers of the world. And of particular interest, to care about their welfare: to fly into action when they are distressed or in danger, and to feel pleasure in their contentment. Human mothers share these urges with every mammalian mother alive today—all thanks to our cynodont ancestors.

I am a mother myself, twice over, and I am all too aware of how heady and overpowering maternal love can be. That said, the births of my daughters did not represent the first or only times I have experienced fierce, heady, gripping love. You may recognize many of the features of maternal love from other loving relationships you've had—romantic love in particular has many of the same intoxicating qualities. There is a reason for that. It is thought that the capacity for love and caring of all kinds, from romantic to filial to sibling love, even love of friends or pets, grew out of the capacity for maternal love. Once the proto-mammalian brain was equipped with the wholly novel and evolutionarily necessary capacity to care about the welfare of other beings outside the self, there was no limit to what other kinds of love could theoretically be felt. It's little wonder that the ethologist Irenäus Eibl-Eibesfeldt viewed the emergence of maternal nurturing as "a turning point in the evolution of vertebrate behavior—one of those celestial moments that [a poet] would call a star hour."

I don't think that is an understatement. As far as we know, the vast, cold universe existed for billions of years without love existing anywhere within it. Then, following the need of a few furry proto-mammals to keep their young fed and warm, it did. It was an explosion into being as magnificent as the birth of any star.

One of maternal love's most impressive properties is its capacity for spontaneous combustion. Ideally, a burst of love goes off like gunpowder inside every new mother's brain when she gets her first look at or whiff of her offspring and feels their first fumbling efforts to feed. The love needs to come on rapidly and powerfully in part because it needs to override or eliminate any fear she might otherwise have of the strange new creatures before her. Altricial babies need food and warmth right away—there is no time for a protracted warming-up period.

If it strikes you as silly to be afraid of a baby, that's only because you are the descendant of cynodonts. Objectively speaking, altricial mammalian babies look funny and smell funny and make horrible, strange noises, and to top it all off, they are by definition complete strangers. Their mother has never encountered them before in her life. Normally, a little healthy fear of anything strange and new is a good thing. But a mammalian mother can't be put off by her babies' funny looks or smells or newness. Her love must draw her near to these little strangers immediately and keep her there, giving them food and warmth despite everything. *And,* as long as she's going to commit all these resources to her babies, it would really be ideal if her babies survived to make her investment worthwhile. So she needs to not only stay physically close but to pay close attention to the welfare of each baby too. Is it content? Or is it upset? Does it need something? More food? Warmth? Did it somehow stray from the nest or lose track of her? Is it trying to get back to safety? Does it need cleaning? Is it hurt? Is it in danger?

That all of these new features—altricial young, milk production, attentive maternal care, and intense investment in each offspring—came bundled together might seem improbable bordering on miraculous. But this bundling is in perfect accordance with something called *life history theory*. According to the theory, the reproductive strategies that a given species can adopt lie on a continuum. At one end of the continuum sit species like loggerhead turtles that are *r-selected,* meaning that their reproduction is limited by the available resources. R-selected species tend to produce precocial offspring, invest few resources in them, and provide them with little or no care. As a result, most of these offspring die young. But r-selected species produce so many offspring that only a few need to survive to keep the species going. It's a mass-production approach to reproduction that seems terribly heartless and wasteful to us because our ancestors shifted so dramatically away from it. Cynodonts and most of their descendants follow more *K-selected* strategies, which you could think of as an artisanal approach to reproduction. Mothers who follow a K–selected strategy produce altricial offspring and so must devote time and energy to keeping them alive. As a result, they can't have nearly as many of them. Instead, they lavish the few babies they have with nourishment and care to maximize their odds of surviving to adulthood. We humans, with our small numbers of desperately needy and altricial offspring who require well over a decade of parental time and energy before they reach maturity, are about as K-selected as an animal can get.

The changes in reproductive strategies that our small, warm-blooded ancestors underwent to keep their babies alive beautifully capture the reason why a mother loggerhead could lay her eggs in the sand and then disappear without giving the progeny they contain a further thought while I found myself overwhelmed by anxiety and protective feelings for her hatchlings. Her chelonioidea brain simply could not conceive of love. It doesn't have the requisite wiring, because it never needed to. But my K-selected mammalian brain comes prepared to love and dote on babies, to treat each one like a precious gem that requires abundant care and nurturing and protection.

"Okay," you might reasonably retort, "that would make sense if you were talking about caring for your own babies. But why on earth would your K-selected mammalian brain have prepared you to care for *turtle* babies?"

This is not a bad question.

In truth, many mammals would not care at all about turtle babies. They would be much more likely to step over them or on them, or to eat them, than to usher them to the water or heckle gulls on their behalf. These differences stem in part from how precisely tuned a species' parental nurturing response is. All mammalian species must come prepared to care for their own babies. But the degree to which they will care for *other* babies varies widely. Ruminants like sheep, for example, are usually uninterested in any babies other than their own. After a newborn lamb first drops to the ground in an amniotic clump, his mother will spend the next several minutes assiduously licking him off, then nudge him to begin nursing. Their licking and nudging and nursing allows the mother and lamb to learn each other's unique smells and to imprint upon one another. After a few hours, the imprinting window will close, and it is very difficult to get it to open again. Thereafter, the ewe will reserve all of her nurturing and milk for her own lamb and skirt away from or even butt away any other lambs who try to intrude. If a lamb is orphaned, it is unlikely to be adopted and will probably starve, even in a flock full of bulging udders. You can bet that a ewe who would complacently allow a herd-mate's lamb to starve to death would not give a fig about a turtle hatchling.

Compare this to the open-armed mothering that can be observed in the humble rat. Rats are devoted mothers who will work strenuously to stay close to their pups. They will cross painful electrified metal grids—the rat equivalent of walking over coals—if their babies are marooned on the other side. Mother rats will tolerate more such suffering to reach their pups than hungry rats will undergo to reach food, or thirsty rats to reach water. But perhaps more remarkably, rats will struggle and suffer even for babies that are not their own, even for babies they have never seen before.

My colleague Stephanie Preston, a psychologist at the University of Michigan, recently unearthed a long-forgotten 1968 study by William Wilsoncroft that demonstrated this fact. Wilsoncroft and his students selected five pregnant rats for their study and taught each one that when she pressed down on a bar in her testing box, a piece of Purina rat chow would come tumbling down a nearby chute into her dish. If pregnant rats experience the same rapacious hunger I did when I was pregnant, I'm sure they were all too happy to learn this task.

Then, after the mother rats gave birth, they were given one day to bond with and nurse their pups before the researchers whisked the pups from their nests. When the mystified mothers emerged from their empty nests to return to bar-pressing, they once again found that chow came tumbling down the chute. At least, it did after the first six bar presses. The seventh time a mother pressed the bar it was followed, not by the dry rattle of a piece of rat chow, but the soft thumping of a pink, hairless pup. Down the chute tumbled one of the mother's lost pups, right into her dish! I leave it to you to imagine what the mothers made of this development. Two of the five immediately did what attentive mother rats do, which was to pick up the pup gently in her teeth and carry it back to the safety of the nest, some three feet away. The other three pressed the bar a few more times first and found that after each press another pup tumbled down. Soon all the mothers were on the same page—pressing the bar and toting their lost pups to their nest.

But the experiment wasn't over. On the thirteenth bar press, something new happened again: this time the pup who tumbled down the chute and into the cup wasn't one of the mother's own pups but a stranger. She had never encountered this pup before in her life. Now what? If mother rats were like ewes, the strange pups would be doomed to starve in the dish. Fortunately, mother rats take a different approach to mothering, which is a pup is a pup is a pup. None of the five so much as hesitated. They all picked up the strange pups, carried them back to the nest, and deposited them among their own pups. Then they went back to the bar and pressed again—and again and again, bringing pup after pup back to the nest, both their own pups and pups that were strangers.

How long do you think this process went on? Ten minutes? An hour? The surprising answer is that it never did end. The ones who put a stop to the parade of pup-toting were the *researchers,* who after three hours of refilling the hopper with pups ended the experiment in exhaustion. During this time the supermom of the group, Rat 5, retrieved an astonishing 684 pups from the dish (although in reality it was the same set of 20 or so pups, recycled many times over). That's a pup every fifteen seconds for three hours, without a break, carted back to the nest for a total distance of over 2,000 feet. But even the *least* energetic of the mothers retrieved a total of 247 pups back to the nest. Moreover, all the mothers were equally attentive when confronted with both their own and strange pups, which suggests that for some mammalian mothers, anything that trips the "baby!" alarm is a call to maternal action (even if, again, this alarm did not always sound equally urgently in all the mothers).

This take-all-comers approach to mothering is not unique to rats. Many other group-living mammals excel in *allomothering,* which literally means "other mothering," or taking care of infants other than one's own. Allomothering can include anything from toting around other mothers' babies to protecting them from danger to nursing them. In some cases, allomothers even foster or wholly adopt orphaned animals. Allomothers are often sisters, aunts, and other adolescent or adult female relatives, but not always. Females of many species allomother unrelated infants as well. And despite what the term implies, males also allomother, contributing to nearly every aspect of parenting other than nursing across a variety of species.

The emergence of allomothering within a species is largely driven by the neediness of the species' young. It's a behavior that is at least three times more likely to arise in animals that bear altricial babies. Hence sheep, which are born woolly and able to stand and walk within minutes, do not allomother, whereas rats, which are born naked, blind, and totally helpless, do. Among the many other mammals renowned for their energetic allomothering are meerkats, seals, sea lions, jackals, wolves, domestic dogs, and lions. Consider for a moment the fact that these highly nurturant species are also all predatory carnivores. Their ferocity on the hunt makes for a striking juxtaposition with the gentleness and devotion that female jackals, wolves, and lions show not only their own but other females' young. The same lioness who runs down and eviscerates a wildebeest one minute may well groom and nurse another female's cubs the next. It also makes for an interesting contrast with sheep, who are often depicted as quintessentially innocent and gentle creatures but can be callous or downright cruel toward each other's lambs. It's a good reminder that there is nothing unnatural or even surprising about the capacities for genuine ferocity and genuine nurturing care coexisting within a single species, or a single individual.

One spectacular demonstration of this is a Kenyan lioness dubbed Kamunyak, or "Blessed One," who in 2001 was discovered in close proximity to an oryx calf that, to the astonishment of the locals who discovered them, was not dead. Oryxes are antelopes, which are favored prey for lions. So what was going on with Kamunyak? Nobody knows for certain, but it is thought that she may have chanced upon the calf and its mother, and the mother had bolted. Ordinarily that would have been the end of the calf. But not only did Kamunyak not attack the abandoned calf, she *adopted* it. For days she could be seen lying or walking by its side, occasionally even grooming it with her rough tongue. She responded attentively when it cried out and was fiercely protective of its safety, keeping it away from humans who ventured too close and at one point chasing off a leopard. She wouldn't leave the calf long enough to find food for herself, so she must have been very hungry. Yet still she cared for the little calf until it ultimately met a sad end—it was killed and eaten by another lion after it ventured out of Kamunyak's sight for a moment. By all appearances, Kamunyak grieved the death of her charge, rushing the lion who had attacked it and then, when nothing more could be done, wandering around roaring mournfully. But it wasn't long before Kamunyak was spotted with another calf—the second of six oryxes that she would ultimately attempt to adopt. Her mothering skills improved over time; starting with her third adoptee, she briefly allowed the calves to nurse from their mothers before chasing the mothers away again.

Although her behavior was undoubtedly unusual—visitors to her park flocked to see a lion quite literally lying down with, if not a lamb, a similarly vulnerable antelope calf—Kamunyak was no singularity. Another lioness in Uganda was recently seen toting around an antelope calf by the scruff of its neck as though it were an unusually gangly cub; later she was spotted allowing it to try (fruitlessly) to nurse. And in 2014, photographers in Botswana observed yet another lioness kill an adult female baboon that was carrying a baby at the time. The infant tried to escape but wasn't yet strong enough to climb a nearby tree. The lioness walked over to investigate and, spotting the baby, pawed at it gently with her massive forefoot. Eventually, she picked the baby up—as though it was one of Wilsoncroft's tumbling rat pups—and carried it a short distance away, whereupon she lay down in the shade and deposited it between her front feet. The photographers captured images of the baby trying to suckle from the lioness's bristly chest, and of the lioness charging a male who attempted to encroach. Retrieval, caressing, nursing, defense—the full suite of mammalian mothering behaviors—were all shown by a lion for a baby baboon!

Despite how bizarre it might seem for a wild lion to tenderly nuzzle and care for a baby antelope or baboon, this behavior is not difficult to explain in the context of allomothering. To effectively allomother each other's young, lions must maintain a relatively relaxed threshold for what sets off their "baby!" alarm and triggers their urge to provide protection, affection, and care. And just as is true for rats, both the threshold and sensitivity of the "baby!" alarm and the energy with which maternal care is provided inevitably vary among individuals, just as surely as individuals vary in size and coat color and countless other individual traits. Extreme variation in maternal sensitivity appears to lead some unusually motherly lions to adopt baby oryxes and baboons, both of which do resemble lion cubs in many respects: they're of similar size and coloration and have infantile features that are common to most species, like large eyes and foreheads and small noses and jaws.

Although perhaps only the very most motherly lionesses engage in cross-species adoption, this behavior is near-normal for species that are even more allomotherly, like domestic dogs. Stories abound of dogs who have of their own accord taken up the challenge of mothering infants of diverse species, including those that are dogs' natural predators or prey. One of my favorites is Mimi, a grizzled ten-year-old female chihuahua owned by a Florida woman named Jeanette Young. In 2007, Young took in a litter of four squirrel pups that her son-in-law had found in a downed nest. Initially she didn't let Mimi near them, which seemed reasonable—there may be no wild animal that elicits more crazed canine hunting behavior than squirrels. But, as Young told CNN, Mimi became progressively more obsessed with the pups, continually approaching them and whining and, as Young put it, "carrying on." So finally, although still somewhat fearful that Mimi might try to attack or eat the pups, Young put the nest on the ground to see what Mimi would do. To her surprise, Mimi almost immediately adopted the pups as her own. Right away she began licking them vigorously, and soon she became protective of them, refusing to venture far from them or to allow Young to approach them. Most spectacularly, after a short while watching over "her" pups, Mimi began producing milk for them and allowing them to nurse. This despite not having given birth to pups of her own for four years!

This is one of roughly a billion stories about domestic dogs caring for animals of other species. In some zoos, it's actually protocol to give orphaned animals like tigers or cheetahs to dogs to raise. An Australian shepherd named Blakely has been designated the "resident nursery companion" at the Cincinnati Zoo, where he (yes, he) has helped to raise dozens of zoo animals that were abandoned or orphaned, including cheetahs, a warthog, and a skunk. His primary job is teaching his charges mammalian social behaviors by gently roughhousing and playing with them. In other zoos, female dogs nurse orphaned baby animals of various species; for instance, a golden retriever named Izzy at a Kansas zoo raised three newborn Bengal tigers along with her own pups. The first time they were introduced, the cubs took immediately to their foster mother, and Izzy to them, and she nursed and nurtured them successfully until their first birthday approached and zoo owners finally separated them. But at no point did either dogs or tigers exhibit hostility toward each other. On what would be their last day together, photographs show Izzy nuzzling the face of one of the 140-pound tigers she raised. Many other examples abound of domestic dogs nursing or otherwise caring for infant tigers, lions, cheetahs, red pandas, deer, wild dogs, pigs, ducks, owls—you name it. There seems to be no limit to the kinds of animals that dogs will allomother.

Although carnivores like lions and dogs can be surprisingly good allomothers, primates are, as a group, even better. From tiny tamarins and marmosets to siamangs, many primates are wonderful allomothers (although, interestingly, great apes like chimpanzees and orangutans allomother relatively little).

But the real allomothering superstars are humans. As the anthropologist Sarah Hrdy describes in her classic book *Mothers and Others,* we humans owe our survival as a species to the vigor and promiscuity with which we nurture each other's children. Whereas other great ape mothers are jealously possessive of their newborns, human mothers traditionally have sought and received sustained help with their babies right from the start. Hrdy and other anthropologists have studied members of modern foraging societies throughout Africa, Asia, and South America for clues as to how our ancestors most likely lived—and hence how our species evolved. They have found that at the *low* end of the human allomothering spectrum are the !Kung of southern Africa, among whom infants are in the care of someone other than their mother roughly one-quarter of the time. In other societies, like the Hadza of Tanzania, babies are cared for by someone other than their mother a whopping 85 percent of the time during their first days of life. Although mothers take over more as time goes on, Hadza children still spend roughly one-third of their infancy being carried by allomothers. Central African Aka and Efe women share their babies, cooperatively holding, comforting, washing, and even nursing them collectively. This is no anomaly; women provide children who are not their own with milk in nearly 90 percent of modern foraging societies around the world.

Humans in modern cultures extensively allomother as well, although allomothering takes different forms across various cultures and subcultures. Most modern human infants don't receive care from twenty different people *every day* like the average Aka baby, but they will nonetheless be allomothered by a great many adults and older children between infancy and adulthood. This includes all the feeding, cleaning, cuddling, protection, and entertainment that infants receive from their earliest days onward from fathers, siblings, grandparents, aunts and uncles, and cousins—as well as from genetically unrelated caregivers like doctors, nurses, neighbors, and babysitters. It also includes the care, instruction, and resources that older children receive from teachers, coaches, and other adults as they mature. And of course, children who are raised by adoptive parents, foster parents, or godparents are, biologically speaking, allomothered as well.

I sometimes get the sense that some people view it as a necessary but unnatural evil that human children receive so much care from individuals other than their mothers. The belief that children should spend the maximum amount of time possible specifically with their mothers for proper socioemotional development is common and not entirely without basis in the scientific literature. Developmental scientists tracing back to the psychiatrist John Bowlby have historically emphasized the importance of a secure attachment to a single primary caregiver (nearly always the mother) for a child's social and emotional well-being. Given this, the assumption that a child will inevitably be at least a little worse off when cared for by anyone other than his or her mother is common. This assumption underlies many public debates about whether child care for working mothers should be affordable and accessible, and it probably influences the thinking of the 60 percent of Americans who continue to believe that children are better off if their mother stays home to care for them. Mothers who are not working get, if anything, even more flak if they leave their children in others' care. I was on maternity leave after having my second daughter when I mentioned to an older relative that we'd hired a night nurse to help care for her. His somewhat skeptical response was to ask why I wasn't doing it "the old-fashioned way."

The idea that it is preferable for a human mother to be her child's sole caregiver around the clock is rooted in modern postwar views of family life that have drifted far afield from our species' evolutionary roots. Mothers getting ample help from experienced caregivers is the *actual* old-fashioned way. This way is vastly more natural and sustainable—and beneficial for children—than for one or even two frazzled and inexperienced parents to try to manage twenty-four hours of daily care for some of the most altricial newborns on earth. As the historian Stephanie Coontz has put it, "Children do best in societies where childrearing is considered too important to be left entirely to parents." Allomothering not only relieves mothers of the massive burden of caring for and rearing needy, altricial, resource-hungry children alone but also fosters strong bonds between children and a wide array of supportive adults and provides children with opportunities to learn the many skills they will need as they mature—not least among them the opportunity to learn to love and trust widely rather than narrowly.

I should mention, by the way, that in focusing on mothering and allomothering, I by no means intend to give human fathers short shrift—far from it. Most mammalian fathers play little or no direct role in raising their children, but human fathers are among the more devoted and indispensable "allomothers" in the animal world. Between fathers' direct care for their children and the indirect care they provide for mothers while mothers nurse and tend to children, it is clear that humans' survival as a species depends, without question, on fathers' committed care.

Solo care of human infants is essentially impossible. It's so difficult that mothers in some foraging cultures (as well as mothers of other allomothering species) will abandon their newborns if they perceive that they will not receive sufficient allomothering support. The prevalence of postpartum depression has much less to do with postnatal hormones (a common myth) than with how legitimately depressing it is to care for a baby without enough help. Inadequate social support is a top risk factor for postpartum depression; it's a bigger risk factor than poverty or having medical complications. My own experience bears this out. My husband and I tried to take care of our first daughter without enough help, and the effects on our mental health were grim. We were not living in poverty, and I suffered no complications, but we were grossly inexperienced at taking care of babies, had no close family living nearby, and were miserable. We were much smarter about paying for extra allomothering the second time around. The salary we paid Marie, our lovely night nurse, to help care for my second daughter may be the best investment I've ever made, and it opened my eyes to the importance of allomothering in raising human children.

None of this negates the importance of close bonds between mothers and their own children, of course. Allomothers need not prioritize the welfare of all infants equally, or be equally invested in their care. But humans' deeply ingrained capacity for allomothering explains why most adults find it so difficult to disengage from the sight or sound of children in need, even when those children are total strangers. This is why charities often use images of distressed children in advertisements to such powerful effect. Depictions of suffering or injured children can galvanize action from strangers in a way that hordes of suffering adults rarely do. Think of the famous "napalm girl" photograph in which a naked and burned nine-year-old Phan Thi Kim Phúc and several other children are running, terrified and screaming, from a napalm attack on their village. This searing image was credited with turning the tide of public opinion against the war in Vietnam. Think also of the awful, heartrending image of Aylan Kurdi, the Syrian toddler whose sweet, round body was found facedown and lifeless amid the lapping waves on a Turkish beach. The sight of this one boy stirred compassion in millions of strangers and was widely credited with causing large shifts in opinion about Syrian refugees. It also massively increased charitable donations to help them. One charity, the Migrant Offshore Aid Station, which rescues refugees' boats in the Mediterranean, saw a fifteen-fold increase in donations in the twenty-four hours after the photograph was published. The average number of donations to the Swedish Red Cross increased *one-hundred-fold* in the week following the photograph's publication relative to the week immediately before.

That intensive allomothering is a fundamental part of human nature helps to explain why we humans have unusually low and relaxed thresholds for what will trip our "baby!" alarms. We put lions and dogs to shame. But our alarms work fundamentally the same way as theirs: the urge to care is tripped by the perception of what ethologists call "key stimuli" that characterize babies and young children. These features include a large head, large eyes, and a small chin and jaw, features that set children apart from adults across nearly every vertebrate species and that elicit care from adults very effectively. The reason that babies look this way is fairly straightforward: their brains and the tops of their skulls grow fast and early, whereas the lower halves of their faces grow slowly. These patterns are particularly pronounced among mammals, in whom milk supports the growth of babies' ballooning craniums. The resulting babyish proportions, termed *kindchenschema* by the ethologist Konrad Lorenz, create an appearance of cuteness and cuddliness that draws in adults' attention and causes them to respond with increased care.

For example, Jim Coan and his colleagues at the University of Virginia found that after adult men and women simply view pictures of baby animals, they make slower, more careful body movements. This is particularly true when the images depict unusually cute, babyish-looking animals. My own research has found that viewing images of human infants triggers the motivation to approach—to literally draw closer. With my student Jennifer Hammer, I presented forty-five subjects with images of unfamiliar adults and infants on a computer screen and asked them to alternately categorize each picture by moving a lever toward or away from themselves. The goal of the study was to measure the degree to which different images caused respondents to either want to approach what they saw—to draw closer—or to avoid it. Approach and avoidance are among the most primitive emotional responses that we have, and nearly any emotionally significant stimulus will trigger the desire to do one or the other (or occasionally both). Approach and avoidance can be measured in the laboratory using a lever, which in our case was a joystick hooked up to a computer. When we compared how quickly subjects pushed or pulled the lever in response to each face, we found that they pulled and pushed equally rapidly in response to adult faces, meaning that they felt no particular motivation to either avoid or approach. But when subjects (both men and women) gazed at babies' faces, they were suddenly able to pull the lever toward themselves—indicating the motivation to approach—much more quickly than they could push it away. We also found that this pattern was linked to psychopathy. Subjects who were less psychopathic—in other words, more caring—had stronger approach responses to babies, suggesting an intrinsic link between this response and compassion.

The psychologist Leslie Zebrowitz and others have documented many other ways in which a cute, appealing, babyish appearance elicits caring responses from adults, whether the possessor of the babyish appearance is an actual baby or an adult with a babyish-looking face, or even a babyish-sounding voice. Adults with babyish features are viewed as more worthy of concern and care and as needing more of it. They receive lighter judicial sentences for minor crimes and are more likely to receive help from strangers—in one study, adults with babyish-looking faces were more likely to have lost résumés mailed back to them than were less babyish-looking people. These effects are not simply due to babyish-looking people being more attractive. Even controlling for variation in attractiveness, more babyish-looking or -sounding people evoke more concern and protectiveness from others who encounter them. Note that the urge to nurture and care for that which is babyish doesn't stop with human adults but extends to babyish-looking cartoon characters, toys, and nonhuman animals, all of which attract attention, approach, and the urge to care.

This fact helps to explain all the abundant alloparental care that humans provide to nonhuman animals as well. What do you think domestic dogs and cats are but unusually infantile-looking wolves and wildcats that people provide with the full suite of maternal care, including retrieval, cleaning, feeding, and protection? Over half of all American households at any given time routinely engage in these behaviors, a fact that researchers have struggled to explain. There appears to be no genuinely rational reason that so many people across a wide variety of cultures keep and care for pets. Although some of them fulfill nominally useful functions like protection or catching vermin, household pets, particularly those in industrialized countries, are mostly an enormous net drain on time and resources, and they require constant care—not unlike human babies. It may be that we just can't help ourselves. We are literally built to parent, and we have fairly lax thresholds regarding who or what we will apply our allomothering energy to. It has been speculated that one of the causes of rising pet ownership in industrialized nations is the decline in these nations' fertility. It's as though we seek out pets to satisfy allomaternal urges that aren't met by human babies.

Humans are also responsive allomothers to wild baby animals of all stripes. Just think back to the crowd on the Florida beach where I shepherded the baby loggerheads to the sea. That day I was part of a glorious assemblage of men and women and boys and girls all allomothering our hearts out together. Those turtles were just a few of the thousands upon thousands of wild animal babies helped or rescued by Americans every year—like the squirrel pups adopted by Mimi the chihuahua, who would never have met their new allomother had they not first been retrieved, protected, and fed by two humans, Jeanette Young and her son-in-law. Even in an urban metropolis like Washington, DC, wildlife allomothers abound. I became one of them yet again shortly after setting out for a run through my neighborhood one recent September.

I had just turned on to a large, busy boulevard when I encountered a man and a woman on the sidewalk staring up into a tree. I paused long enough to spot tufts of gray, fluffy feathers and a slate-gray beak poking out between the curled fingers of the man's hand. My curiosity got the better of me, and I stopped to ask what was going on. The man held out his fist to show me a baby blue jay cupped inside. It was quite still, but alert and unharmed; it gazed at me with glittering black eyes. The man told me he had been driving by and saw it hunched in the middle of the road, still wearing the downy feathers of a nestling and unable to fly. He pulled over, leapt from his car, and ran into the street, evading squealing rush hour traffic, to retrieve it and carry it to safety. But now he didn't have the faintest idea what to do with it; he and the woman had been trying to locate a nest in the tree above. But they were both on their way to work and couldn't stand there indefinitely. Isn't it interesting that none of the three of us ever even considered simply leaving the little creature on the sidewalk to fend for itself? Unthinkable. What could I do but offer to take it? So I did. Gently, the man transferred the bird's warm, cloud-soft body into my hand, then thanked me and drove away. I carried the jay home, its scaly dinosaur toes clutching my fingers and its heart pounding against them. It never made a sound, and its eyes never left my face. Once home, I put it inside a small box lined with a cloth, then drove it to a local shelter called City Wildlife, where it was raised with foster siblings brought in by other local allomothers before being released back into the urban forest. I think of it sometimes still. I hope it is doing well.

Surely you see where I am going with all of this.

Allomothering, particularly the forms that involve retrieval and protection, is all but indistinguishable from extraordinary altruism. What my heroic rescuer did for me in 1997, an anonymous neighbor here in Washington, DC, did (in an only slightly less extreme form) for a baby blue jay: he spotted a vulnerable creature in distress, felt an instantaneous urge to help, then risked his own safety by swerving to the roadside and running into traffic to rescue it. An even more comparable animal example can be seen in a recent video I encountered in which a Russian motorist risked his life to rescue a kitten stranded on a busy freeway. The urge to respond with protection and care to the young and the vulnerable, even at significant risk and cost to ourselves, is our birthright as mammals. The urge to respond this way even to youngsters who are not our own is our birthright as group-living social mammals who bear altricial young. But a birthright we share with only a very few other species (domesticated dogs among them) is the urge to respond to creatures of a wide range of ages and species—even if we have never seen them before, even if we might in other circumstances consider them predators or prey or pests—if they manage to trigger our highly sensitive and generous "baby!" alarms. It is this urge that lays the groundwork for the emergence of extraordinary altruism.

Allomothering *is* altruism, really. Species that allomother are perennially attuned to vulnerability, distress, and need, and they are primed to respond with nurturing and care when they spot it, even if the object of their care is unfamiliar or unrelated to them. A rat pup rescued by a strange adult female, an oryx calf protected from leopards by a lioness, or a blue jay saved from the street owes its life to the person who rescued it just as surely as I owe my life to my roadside rescuer, or Priscilla Tirado owes her life to Lenny Skutnik, or Zina Williams owes her life to Cory Booker, or thousands of patients in kidney failure have owed their lives to anonymous donors. And it may be exactly the same neural mechanisms that impel all of these behaviors.

Direct evidence that allomothering provides the basis for altruism came from a recent study of humans and other primates that was conducted by Carel van Schaik at the University of Zurich. He and his colleagues were seeking to identify evolutionary causes of altruism, or what they termed *proactive prosociality*, in which one individual spontaneously helps another without receiving any gains in return. They tested the tendency of twenty-four distinct species of primates, including lemurs, monkeys, apes, and humans, to engage in such behavior using a simple task in which one individual could work to provide other individuals with food while gaining nothing personally. The researchers then examined what other factors were associated with rates of helping across the various species. They considered factors like brain size (a proxy for intelligence), overall social tolerance, frequency of cooperative hunting, and whether the species tends to form strong pair bonds. They also examined rates of allomothering. They found that across all the primate species, including humans, the *single best predictor* of altruistic behavior was allomothering. Among macaques and chimpanzees, who allomother very little, altruism among adults was almost nonexistent. Among tamarins and humans, who have very little in common except extensive allomothering, altruism was frequent. Most of the other variables the researchers considered ceased to be related to altruism at all once allomothering was entered into their statistical model. The moral: species who care for one another's babies are also far more likely to help each other out—to be altruistic—even when they don't stand to gain anything from it. The researchers concluded that "the adoption of extensive allomaternal care by our hominin ancestors thus provides the most parsimonious explanation for the origin of human hyper-cooperation"—or, in other words, altruism.

This explanation makes a lot of sense. The evolution of mothering is widely agreed to be the origin of the capacity to care about the welfare of any being outside the self; the evolution of allomothering represents the origin of the capacity to spread that care far and wide rather than hoarding it greedily for one's own progeny.

But can the relationship between allomothering and altruism also help us understand the rarer phenomenon of extraordinary altruism in humans? I believe it can.

In all the species I've described, individual variation in allomothering responsiveness is evident. Both the impulse to care for infants and the threshold for what triggers the urge to care vary considerably, not only across species but within each species as well. Wilsoncroft studied only five mother rats (who were all first-time mothers), but found clear variation in their mothering ability and motivation. Most lions kill any baby antelopes or baboons they encounter, but a few do not, and some, like Kamunyak, even go to great lengths to care for and protect them. And of course, humans also vary considerably in their allomothering interest and ability, as well as in their altruistic tendencies. The key question is this: what, if any, direct evidence is there that variation in allomaternal responsiveness underlies extreme acts of human altruism? I have already described it to you.

Recall that my own research, and that of others, finds that one of the best predictors of altruism is responsiveness to fearful facial expressions. Individuals at the very low end of the caring continuum—psychopaths—are notably insensitive to these expressions, probably as a result of dysfunction in the amygdala. They fail to recognize fearful expressions and fail to show appropriate emotional or behavioral reactions to them. Whereas fearful expressions seem to inhibit aggression and elicit empathic concern in most viewers, people who are psychopathic are relatively impervious to their effects. Altruists, on the other hand, are unusually sensitive to these expressions. They recognize them better and show enhanced emotional reactivity to them.

The reason this is so incredibly interesting is that, of all the expressions that a human being can make, the one that reconfigures the face to most resemble that of a baby is fear.

Fearful eyes are wide and large, just like a baby's eyes, the visible portion of which has already attained adult size by three months of age. Fearful brows are high and angled upward, while the mouth is rounded and low and the jaw is small and receding. Together, these features make fearful faces appear vulnerable, submissive, appeasing, and infantile. Without a doubt, if you were trying to maximize an adult human face's resemblance to a baby, you'd make it look fearful. That fearful expressions *do* look babyish has been empirically demonstrated. Some time ago, I published a paper with my undergraduate adviser Robert Kleck and my colleague Reginald Adams Jr. that showed that adopting a fearful expression causes a face to appear more babyish in every sense. Viewers described an array of fearful faces we showed them as infantile and dependent and as possessing all the "key stimuli" of actual babies' faces, including large eyes, high brows, a small jaw, and a rounded appearance. They perceived fearful faces as babyish even when the faces were altered to retain their key appearance features but to be no longer recognizable as expressing fear, demonstrating that it is the physical appearance of the expression that makes the expresser appear babyish. This may explain why, as my student Jennifer Hammer and I recently found, people respond to fearful expressions with the same patterns of approach that they show infants' faces, and why this urge to approach is also

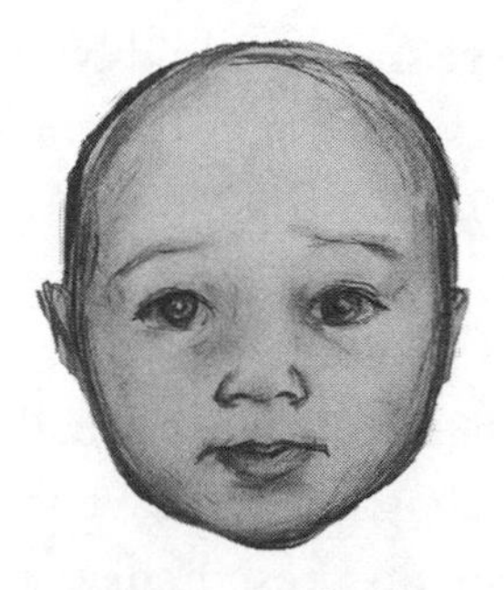
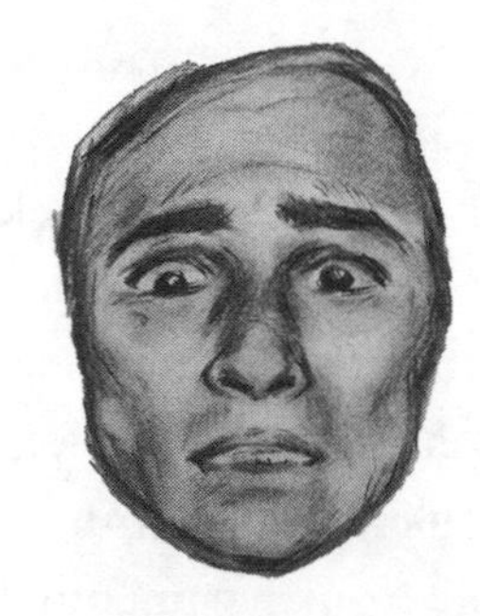

The fearful facial expression, which incorporates enlarged eyes, high oblique brows, and a rounded mouth and jaw, causes an adult's face to more closely resemble the face of an infant. *Abigail Marsh.*

strongest among people who are the most compassionate (and the least psychopathic). The similarity in this study between how people respond to babies' faces and how they respond to fearful expressions was striking.

That fearful expressions evolved to look the particular way they do is almost certain. These expressions, like happiness, anger, and other basic facial expressions, are displayed and recognized by members of many cultures around the world. A meta-analysis conducted by my graduate mentor Nalini Ambady and my colleague Hillary Anger Elfenbein made this clear: they found that in hundreds of studies conducted across dozens of cultural groups around the world, viewers can reliably interpret the meaning of fearful, angry, happy, and other expressions displayed by members of even distant cultures. Considering that some of our nearest primate relatives also express their fear, anger, and happiness using similar facial behaviors, it is very unlikely that these expressions are a purely culturally learned or socialized behavior. Instead, these expressions may serve vital functions that have caused them to be evolutionarily conserved across the generations.

The most powerful evolutionary explanations apply across multiple species—much as van Shaik and his colleagues were able to capture the relationship between allomothering and altruism by looking across multiple species of primates. So it is useful to consider the functions that fearful behaviors serve in other highly allomaternal species, like dogs and wolves. Dogs and wolves employ distinctive behaviors when fearful of attack. They crouch down or roll over, fold their legs and tail close to their bodies, and flatten their ears. They may whimper, lick the jaws of the aggressor, or even urinate. Together, as explained earlier, these behaviors usually trigger the observer's Violence Inhibition Mechanism and save the cringing, crouching wolf from harm. But *why* do these behaviors inhibit violence? Or, to alter the emphasis slightly, why do *these* behaviors inhibit violence? Because, together, they cause the fearful wolf or dog to take on the appearance and other key traits of the one creature that social mammals who bear altricial young are very unlikely to attack: a baby. Key stimuli that set wolf pups apart from their parents are their small size, supine posture, and flattened ears. Pups also emit high-pitched cries, lick their parents' jaws to request food, and sometimes pee on themselves. *That* is how these cues inhibit violence: in combination, they very effectively trip not just the aggressor's "baby!" alarm but the even louder and clangier "oh my God, *a baby is in trouble!*" alarm, which rapidly suppresses the urge to attack and replaces it with care.

Human fearful expressions, it seems, may do precisely the same thing—as do other fearful cues, like crouching, frightened body postures, or shrill, high-pitched, fearful screams that echo the high-pitched cries of infants. By signaling both distress and infantile vulnerability, these expressions of fear are tailored with extraordinary precision to move those who encounter them to care. The moral philosopher Adam Smith seems to have intuited this when he wrote: "The plaintive voice of misery, when heard at a distance, will not allow us to be indifferent about the person from whom it comes. As soon as it strikes our ear, it interests us in his fortune, and, if continued, forces us almost involuntarily to fly to his assistance." This description strikes me as uncannily similar to the way in which many altruists I've interviewed described their "almost involuntary" urge to help once they had been alerted to another's misery. Smith's only mistake, perhaps, was to infer that this is a response that is equally strong in everyone. That is almost certainly not the case. Rather, those individuals among us who are the most sensitive to these powerfully care-evoking cues are also—not remotely by coincidence—the most altruistic.

⁎

I am willing to bet that by now you can easily guess what part of the brain is considered the entry point into the parental care system. Yup, it's the amygdala. Of course, the amygdala isn't solely responsible for parental care any more than it is solely responsible for any other cognitive or behavioral outcome. But it is essential for getting parental care going.

Incoming sensory information of any motivational significance gets channeled inexorably toward the amygdala. When key baby stimuli are detected—like the large head and eyes and small lower face that create babies' classically cute appearance—off to the amygdala this information goes. Leslie Zebrowitz and others have found that any face that carries these appearance cues reliably engages the amygdala, regardless of whether the cues belong to an actual baby or an adult who just looks like a baby. Babies' cries are directed to the amygdala as well, and listening to them results in more activation there than do non-cry sounds of even very similar pitch and loudness. This fits in with the amygdala's perennial alertness to signs of distress, like the bright pop of a wide, frightened eye or the ragged sound of a scream. The reason for the amygdala's strong response to fearful cues, then, is likely twofold: fearful expressions and screams not only signal distress but carry infantile features that are important to the amygdala in their own right. The really interesting question is, what happens next? What happens when stimuli like fearful expressions or screams that signal both babyishness and distress arrive in the amygdala together? How does the signal processing in the amygdala lead to the urge to care?

We already know a little bit of the answer, of course. Seeing or hearing distress cues leads to an internal simulation of the distressed state. Externally, this registers as a slight uptick in heart rate, blood pressure, and sweat on the palms. This is an empathic response, in the sense that it helps the person experiencing these changes recognize and understand the other person's distress. But another change that occurs when people see someone in distress is a little counterintuitive. Usually, events that cause mild fear reactions in people, like the sight of a snake or a gun, also trigger the urge to flee, to escape. But we know that, in the average person, this is precisely what people do *not* desire to do in response to others' distress. The lever tasks I've run show that most people respond to others' fear with approach, not escape.

This suggests that something approaching alchemy happens within the amygdala in response to others' distress. Although it enables a viewer to conjure up a little internal simulation of another person's frightened state—to empathize—this simulated fear results in the urge to approach, which is a behavior totally incompatible with escape but quite compatible with providing care and protection. And this response kicks in very rapidly—in roughly a second. The instant urge to approach someone who is frightened and vulnerable almost certainly results from that person's similarity to an eminently approachable infant. In our lever studies, people responded the most quickly to both babies and fearful expressions when they categorized the two kinds of faces together instead of separately, suggesting an implicit association between them. What seems to be happening, then, is that after the sight of a fearful face results in a simulated fear response, somewhere within the brain—likely the amygdala, which regulates approach and avoidance behaviors—the rumbling train of behavior that follows switches tracks in the busy railyard of neurons within it. Because a fearful face carries key infantile stimuli, the train is redirected down an entirely separate track toward caring, protective behaviors. Empathy has been transformed into caring.

But who, or what, is doing this switching? In all likelihood, the switchman responsible for turning social lead into gold is not any single brain structure but rather a brain chemical that changes the activity of multiple brain structures—the amygdala included—simultaneously. This chemical is a neurotransmitter composed of nine linked amino acids produced in only one place on earth, which is the hypothalamus of all living mammals. This molecular alchemist is called oxytocin.

It is difficult to specify exactly when or how the first molecule of oxytocin came into being, but it almost certainly was in the brain of a cynodont, as all the cynodonts' descendants—and only their descendants—produce it. Oxytocin—and its sister hormone vasopressin—probably originally cleaved from an older hormone called vasotocin that fish, reptiles (including sea turtles), amphibians, and birds still produce today, and which differs from oxytocin by only one amino acid.

But what a difference an amino acid makes!

Oxytocin is responsible for two key physiological functions that mammals require to reproduce. The first is stimulating contractions in the smooth muscles of the uterus to get babies out of it. If you or someone you know has had labor induced by a drug called Pitocin, you know how effectively oxytocin brings the walls of the uterus crashing inward. Pitocin is just a Frankenstein form of oxytocin created in a laboratory. One of my two labors was induced with Pitocin, and after the nurse inserted the IV and sent it flowing through my veins, I went from no contractions to wall-clawing labor within two hours. I imagine I would have found it fascinating from a scientific perspective had I not been preoccupied by the sensation that I might be about to explode.

The second key mammalian function that oxytocin performs is enabling nursing. It's not involved in milk production itself, but in making sure the milk can be drunk. Milk that is inside a breast will more or less stay there until a baby latches onto the nipple and starts to suck. That strange drawing sensation is transmitted from the nipple up to the hypothalamus, where it prompts a few small clusters of cells to start churning out oxytocin and relaying it to the nearby pituitary gland. From there, the pituitary releases the oxytocin into the bloodstream, where it filters down to myoepithelial cells in the breast, alerting them to release the milk they contain into the nipple. And voilà, out it flows. This neat little process is called the milk ejection reflex, and it has been keeping baby mammals alive for many millions of years.

Of course, again, the physiological ability to lactate is not all that neat in isolation. Producing milk is only useful if it is accompanied by all the behavioral, cognitive, and emotional changes that allow the young to access it and benefit from it. These changes include the desire to spend lots of time in close proximity to the offspring, a lack of fear of them, and various nurturing and protective behaviors that keep them fed, clean, and safe. This seems like an enormous number of new adaptations for mammalian mothers to have acquired in a short period of time, and indeed it was. But incredibly, they are all supported by oxytocin, the same chemical that gets the babies born and the milk flowing. If milk and maternal care are the defining characteristics of mammals—and they are—then you could say that we mammals owe *everything we are* to this one little clump of nine amino acids. It gives me goose bumps when I really stop to contemplate it.

Oxytocin's importance was initially discovered in studies of rats. You might remember that Wilsoncroft's wonderfully maternal subjects were all first-time mothers. That is an important detail, as rats who have never had babies before act quite differently toward pups. They are worse than ewes, believe it or not. Virgin female rats find the smell and cries of rat pups highly upsetting and generally go out of their way to avoid them. If forced to remain in close proximity to pups, they sometimes attack or even cannibalize them. "Ew," you can almost hear them saying to themselves. "I really, really, really don't like rat pups."

But one thing can turn these callous, cannibalizing monsters into attentive mothers willing to spend hours tirelessly rescuing rat pups from a dish, and it can do so almost instantaneously. That thing is oxytocin.

In the days and hours before a rat gives birth for the first time, neurons that produce oxytocin begin to multiply in her hypothalamus. Receptors for the oxytocin molecule also proliferate throughout her brain, sprouting up where none had been before in regions like the olfactory nucleus, which moderates responses to smells; the hypothalamus; the stria terminalis, a ribbon of fibers connecting the hypothalamus and amygdala; and the amygdala itself. These changes seem to set the stage for caring behavior to flourish.

Up through the 1970s, many efforts had been made to identify the neurotransmitters responsible for caring maternal behavior, and they had largely failed. Estrogen, progesterone, prolactin—all are hormones involved in female reproduction and seemed likely candidates for motherliness, but when they were injected into the brains of virgin rats, their aversion to pups remained unchanged. But when Cort Pedersen and his colleagues injected oxytocin into female rats' brains, their responses to pups were transformed in minutes.

At the start of one experiment, the researchers divided up more than 200 virgin female rats into groups. One group was randomly selected to have an inert saline solution injected into the fluid-filled cavities inside their brains. Then they were placed in the center of a cage that also held three squirming pups arrayed in a triangle, three inches apart from each other. Most of the rats did what virgin female rats usually do, which was to ignore the pups. Fewer than one in five showed signs of maternal behavior—although those that did showed the full suite within the hour, including picking up the pups and putting them into a little pile together, licking them clean, building a nest for them out of any materials handy, crouching over them, and retrieving back to the makeshift nest any pups who wriggled free.

For comparison, the researchers tried injecting a number of other chemicals into other groups of rats, including vasotocin, vasopressin, and estrogen. Still no change. But when they injected oxytocin into the rats' brains, the change was fast and profound. Now nearly three-quarters of the rats started to care assiduously for the strange pups—a 400 percent change over baseline. This showed that oxytocin can singlehandedly promote not just mothering but allomothering, as all the pups were strangers to the female. But again, note that even the powerful oxytocin didn't affect all the rats equally. About 20 percent of the rats given oxytocin still failed to show any maternal behavior, and another 7 percent killed at least one pup—as did about 7 percent of the rats in every group.

A tidal wave of research has since fleshed out the critical roles that oxytocin plays in mothering and allomothering. Oxytocin has been shown to induce maternal (and allomaternal) care in a wide variety of species, including rats, mice, rhesus monkeys, meerkats, goats, and sheep—and thus we can reasonably assume that it similarly affects all the other mammals that haven't yet been tested. Female rats whose brains produce more oxytocin tend to be better mothers. Less devoted mothers get better, though, if you give them more oxytocin. In contrast, they do much worse if you chemically block oxytocin receptors in their brain, which eliminates maternal care almost completely. Oxytocin can induce sheep to begin tenderly caring for unfamiliar lambs, something they would normally never do, thirty seconds after it is injected into their brains. This is not terribly useful information for a rancher, as most sheep ranches are not equipped to perform brain surgery on their livestock. But the knowledge of oxytocin's importance has given ranchers another tool for inducing ewes to accept orphaned lambs, which is—and I know this is gross, but it works—to poke either their hand or a special balloon up into the ewe's birth canal and massage her cervix. Fun, huh? But this stimulation triggers a wave of oxytocin production, much the same way a baby suckling does, and causes the motherly lightbulb in even a sheep's fairly dim brain to flicker to life.

Oxytocin acts throughout the brain, and the specific locales in which it acts vary somewhat from species to species. But across species, the amygdala is a central locus of oxytocin's effects. Oxytocin seems to act in the amygdala to reduce any aversion to the unfamiliar smell, sight, or sounds of an infant, preventing avoidance or aggression and opening the door—or switching the track, if you prefer—for care. Very recent research suggests that this is true in people just as it is in other mammals.

Studying oxytocin's effect on people is more challenging than studying it in rats or sheep. Researchers inject oxytocin directly into these animals' brains because oxytocin is a very big molecule—so big that it is thought to have trouble crossing the blood-brain barrier if it's injected into the bloodstream or swallowed. Researchers usually avoid injecting hormones directly into living human brains, so efforts to investigate oxytocin's effects on humans were stymied for a while. But a simple method discovered in the 1990s solved the problem: squirting oxytocin up the nose as a nasal spray. There it gets absorbed through the thin, porous skin lining the sinuses and into the brain itself.

When I began my postdoctoral work at NIMH, I was dying to set up an intranasal oxytocin study to test how oxytocin affects care-based responses in humans. I started getting the paperwork drawn up in 2004, before almost any human oxytocin research had been published. Unfortunately, the internecine bureaucracy of the NIH prevented me from actually running the study until 2006, by which point a wave of human oxytocin research had gotten under way, accompanied by huge swells of hype. Early studies found that oxytocin increased the amount of money people offered strangers in economic games, or the amount of time they spent focusing on their eyes. Oxytocin was quickly dubbed the "cuddle hormone" and the "love hormone." News articles suggested that car dealerships should pipe it through their HVAC systems to increase sales (really). Psychiatrists speculated that it might cure autism (it doesn't, sadly). The research that generated all this hyperbole ended up getting a fairly bad rap, understandably, and later studies have questioned whether some of these early findings were even true. In truth, oxytocin is not a panacea for making all social interactions more cuddly and lovey. Why should it be? Its essential purpose is supporting the care of vulnerable offspring. (In some species, it has also since been exapted to serve related functions, like pair bonding and social recognition.) Thus, sometimes it will promote sweetness and cuddling and other times it will promote wariness and aggression against intruders—these are all forms of maternal care supported by oxytocin.

In an effort to explore oxytocin's care-related functions in humans, my research assistant Henry Yu and I spent two years asking people to squirt doses of oxytocin or a saline placebo up their noses in the Clinical Center of the NIH, just upstairs from where I was scanning the brains of psychopathic adolescents on the weekends. None of our oxytocin subjects tried to cuddle with us, but their behavior did change in ways that were consistent with increased parental care. In one study we found that a few squirts of oxytocin up each nostril increased subjects' preference for infants' faces, but decreased their preference for unfamiliar adults' faces. This is just what you would expect if oxytocin influences not just parental care but alloparental care: increased preference for unfamiliar babies, whose faces carry key infantile stimuli, and wariness of unfamiliar adults who might do a baby harm. Think of sweet Mimi the chihuahua fiercely defending her pups from her owner once her squirrel babies got her oxytocin production humming, or the lioness who attacked other adult lions when they threatened her baby baboon.

We also found—in keeping with the fact that there is *always* variability—that the degree to which oxytocin increased subjects' preference for babies' faces depended on variation in a gene called OXTR, which affects the activity of oxytocin receptors in the brain. We found that people who carried the "A" version of a particular segment of the OXTR gene preferred babies' faces no matter what we gave them, but that people who carried only the "G" version of this polymorphism preferred infants' faces only after a hit of oxytocin. These findings remind me of Pedersen's rat studies, which found that a few rats are maternal even without extra oxytocin, but that most need an oxytocin boost to become fully maternal. Finally, we found that oxytocin also increased subjects' ability to recognize happy facial expressions, although only happy expressions that were fairly subtle and hard to recognize. This was an interesting result, but several other researchers have since reported results that I find much more interesting: oxytocin has even stronger effects on increasing accuracy for recognizing fear. One study found that oxytocin improved people's ability to recognize fear (and only fear) by about 7 percent. Two more studies, conducted in Israel by Meytal Fischer-Shofty and Simone Shamay-Tsoory and their colleagues, found selective fear-recognition improvements of 13 and 20 percent, respectively.

That oxytocin strongly increases sensitivity to fearful faces—the same faces that psychopathic people fail to recognize, that highly altruistic people recognize with exquisite sensitivity, and that evoke approach and caring in people who see them—is, to me, remarkably clear evidence that oxytocin underlies the power of these vulnerable, infantile expressions to cause those who see them to care.

Now, for this to be true, oxytocin would need to be able to accomplish two things simultaneously. It would need to promote a strong empathic response to distress cues like fearful expressions to enable these expressions to be interpreted. But it also would need to inhibit the urge to avoid or escape in favor of approaching and caring for the fearful. As it happens, the findings reported in a 2016 study of rats strongly suggest that oxytocin can accomplish exactly this delicate balance. When researchers gave oxytocin to rats who were under threat, the rats showed all the usual physiological signs of fear—like elevated heart rates—that are part of an empathic fear response. But the rats didn't show any of the avoidance or freezing that normally accompanies fear. This striking pairing—intact fear physiology but not fear behavior, which would enable an animal to *feel* scared but not *act* scared—was mediated by oxytocin's effects in two separate groups of cells within the central nucleus of the amygdala. These findings help to explain what might otherwise be a puzzling phenomenon in rats: anxious rats make much better mothers, including being braver when defending their pups from harm. Their courage seems to result from their unusually vigorous amygdala response coupled with a rush of oxytocin in this structure when their pups are in danger. I felt a thrill of joy when I first encountered these findings, which represent an essential piece of the strange puzzle of parental care, and of altruism more broadly.

So, although no current technology can directly test this hypothesis in humans, here's what I think is going on. Once a signal arrives in a human amygdala (the basolateral region, specifically) that somebody is frightened, two things happen. First, the basolateral nucleus responds vigorously, reflecting the importance of what it has detected. It then transmits the signal it receives to the central nucleus, which gins up an empathic response. For example, it tells the hypothalamus to increase physiological fear responding—pounding heart and sweaty palms and spiking blood pressure. Simultaneously, the vulnerable, infantile qualities of the expression are also being processed, setting off a wave of increased oxytocin production in the hypothalamus. When the oxytocin reaches the central nucleus of the amygdala, it triggers a response in the huge number of oxytocin-sensitive neurons that populate the lateral part of this nucleus. Neurons here suppress fear-relevant activity in other regions of the amygdala in response. These neurons may signal other cells in the central amygdala to inhibit what would otherwise have been an avoidant fear response in favor of a caring approach. These behaviors are regulated through the amygdala's connections with various other structures in the parental care network, like the striatum and the periaqueductal gray, both of which are densely packed with oxytocin receptors.

The sum total of all this activity is a signal from the parental care system to the rest of the brain that there's a sweet, juicy baby out there in the world who's in trouble and needs your help, so don't be a nervous Nellie—go out and get it!

I'd bet money on this system being fouled up in psychopaths (as well as other systems, I should add—psychopathy almost certainly represents not one single dysfunction but a constellation of them). Dysfunction throughout the amygdala prevents people who are psychopathic from registering others' fear strongly in the first place, and even when they do, their oxytocin system probably isn't set up to generate the urge to care, perhaps owing to observed abnormalities in their OXTR gene or other yet-to-be-discovered causes. In altruists, on the other hand, it's likely that both of these systems are exquisitely sensitive. We know already that altruists' amygdalas are highly sensitive to signs of others' distress. Although so far we have data only about how they respond to fearful facial expressions, I would venture that a much wider array of other signs that a person is vulnerable and distressed—screams, frightened body language, crying, or other forms of helplessness or suffering conveyed verbally—might affect them similarly. Many of the altruistic kidney donors I've worked with say that they first felt the spontaneous urge to donate a kidney when they saw, heard, or read a news story about someone suffering from kidney disease. One read a harrowing Reddit post in which a stranger described what life in kidney failure is like and decided to donate that day. Several have been moved to donate after watching someone close to them suffer on dialysis. Harold says that one of the motivations for his donation was seeing an obituary for a child who had died of blood cancer because no bone marrow donor could be found. Lenny Skutnik was moved to dive into the icy Potomac after hearing the chilling scream of a drowning woman. I've always wondered whether my own rescuer caught a glimpse of my frightened face through the windshield. I'll probably never know. Altruists uniformly seem to get a blast of empathic distress in response to these cues, but perhaps owing to highly responsive oxytocin-producing cells in their hypothalamus, or an unusually high density of oxytocin receptors in certain regions of the amygdala, they don't respond by avoiding or escaping what they've seen. They dive right in with care.

If all this is true (and again, I'm betting it is), it would finally explain the fact that altruists are both sensitive to fear *and* brave in the face of others' distress, and that their bravery is instinctive and intuitive. A complicated chain of events within some of the deepest recesses of their brains cause them to act on some of humankind's most primitive, atavistic urges—urges that trace back to our earliest mammalian ancestors, whose babies needed food and care and would never have survived unless their very babyishness had prompted an unconditional desire to provide care and protection.

A high degree of variation in the specificity and sensitivity of these systems results in a small fraction of the population being so acutely sensitive to others' distress and vulnerability that they will respond with the same conviction and certainty when saving the life of a stranger as an ordinary person might feel only when the life of their own child (or mother) is at stake.

7

CAN WE BE BETTER?

AS SHOULD BE clear by this point, an overwhelming body of scientific data supports the conclusion that human beings are in no way fundamentally selfish or callous. All the neural and cognitive tools needed to experience genuine concern for others' welfare and the desire to help those in distress are part of our birthright as mammals who bear altricial, helpless infants in need of care and protection from both their parents and adults other than their parents. These tools include the ability to detect others' distress, the tendency to feel concern when we do, and the desire to help those in distress, even individuals who are unrelated to us. Naturally, people vary considerably in terms of both concern for others and desire to help them, but it is the rare person who is totally blind to others' distress and completely without concern for their welfare. That gives most of us plenty of innate capacity to build on.

Thus, an obvious question arises: Why can people still be so awful to one another? Why is there violence and hatred and cruelty? Why are some 400,000 people around the world murdered every year? Why did the Holocaust happen? Why can untold millions of suffering refugees not find asylum within the many prosperous nations of the world?

When it comes to all manner of crime, cruelty, and callousness, it is clear that the 1 or 2 percent of true psychopaths among us are disproportionately responsible for much of it. But remember that this fact says nothing about "human nature." In fact, as I've emphasized, the fact that psychopaths are so very different from other people only serves to highlight the average person's capacity for genuine compassion and concern. That said, we clearly cannot blame psychopaths for all the cruelty and violence in the world, or even most of it. Among people incarcerated for violent crimes, for example, only about half are true psychopaths. A nation doesn't invade another country or commit ghastly atrocities because the entire nation is made up of psychopaths. And in daily life, minor acts of cruelty and callousness are too widespread to all be the work of psychopaths. If it's true that nature has built most humans so beautifully for compassion, how can this be?

Part of the answer is that nature has *also* built us beautifully for aggression and violence. There is nothing inherently contradictory about this. Consider the case of the lioness who slaughtered a baboon she intended to eat one minute, then tenderly retrieved and groomed that baboon's baby the next—then savagely chased another lion away from "her" baby a moment later. Or consider the sheep who nurses and dotes on her own lamb one minute before callously butting away the hungry lamb of another ewe. Are these creatures *really* caring, or are they *really* cruel? It is both foolish and unnecessary to come down on one side or the other. Both capacities are equally real. Similarly, the question of whether humans are compassionate *or* cruel can never be answered—we are both. At least, we have the capacity to be. The real question is: when do we express compassion versus cruelty, and why, and to whom? A complete answer to this question requires understanding the essential, inexorable influence of culture on the basic biological processes at hand: how the physical and social environments we inhabit shape our views about, and treatment of, the other beings we encounter during our lives, and how our culture may ultimately enable us to expand our capacity for compassion and altruism.

This chapter delves into four considerations that should be kept in mind as we set out to understand how we might become more altruistic.

1. We are already so much better than we think we are.

It is easy to be misled by attention-grabbing atrocities, but try not to be. The actual numbers are clear: goodness is overwhelmingly common, and kindness is the norm, not the exception. Recall the World Giving Index, which is compiled from the results of massive Gallup polls of thousands of people around the world. The 2016 Index tracked how residents of 140 countries responded to three questions that, together, span a wide range of altruistic behaviors: (1) Have you, in the last month, given money to charity? (2) Have you engaged in volunteer work? (3) Have you given help to a needy stranger? All three forms of altruism can be motivated by a variety of forces, but the third question indexes the kind of generosity most likely to represent a spontaneous, caring response to another's distress or need. This question is aimed at capturing direct acts of altruism like helping a lost stranger find their way, picking up something that was dropped, or giving to a needy person begging for help. Helping a needy stranger also happens to be by far the most common form of generosity in the world, according to the Index: over half of the world's population report helping a needy stranger every month. Donating money and volunteering are also remarkably common. Every month over 1 billion people donate money to a charity. And over 1 billion volunteer their time. *Every month.*

The United States is a more generous country than nearly any other nation on earth, according to these three indices. Across the last five years, it has remained the second-most-generous country in the world. Americans donate hundreds of billions of dollars of their own money annually to charities, and spend over 7 billion hours volunteering to help members of their communities (and this number includes only formal volunteering through charitable organizations). And according to the Index, the United States is a particular standout in giving help to needy strangers. Extrapolating from the Index's results, Americans offer help to hundreds of millions of strangers every year in countless unknown acts of direction-giving, belonging-collecting, change-offering, and the like. Their help also includes forms of altruism not assessed by the World Giving Index, like blood donation. Americans donate over 13 million units of their blood to sick and injured strangers annually, and many of these donations represent spontaneous responses to strangers' suffering and distress. Blood donations reliably surge following publicized tragedies like mass shootings or terrorist attacks. Two months after 9/11, the number of Americans who reported having donated blood had increased by 50 percent. Thousands more Americans undergo painful medical procedures to give strangers the marrow from their very bones every year, and many millions more have volunteered to make these donations if asked. And of course, every year dozens of Carnegie Hero Fund awardees and hundreds of altruistic kidney donors take significant risks to save the lives of strangers.

And these numbers reflect only altruism toward humans. Americans also rescue hundreds of thousands of animals every year. When the National Wildlife Rehabilitators Association conducted a survey of animal rescues in 2007, respondents reported having treated over 64,000 rescued birds, 39,000 mammals, and 2,300 reptiles and amphibians that year alone. Numbers like these mean that, on any given day, hundreds of Americans are rescuing vulnerable, helpless creatures that they encounter. Not, of course, that Americans are alone in this regard. On a webpage that compiles the activities of international wildlife rehabilitation groups, you can scroll through a glorious, seemingly endless list of 153 nations, from Afghanistan to Zimbabwe, where organized groups of altruists are coming to the aid of the animals among them. And if my own experiences are any indicator, many of them are doing so despite personal risks or costs to themselves, and despite the absence of any personal gain.

All of this comports with laboratory findings showing that, when given the opportunity to be generous, most people will be, at least some of the time. The amount of goodness displayed freely and

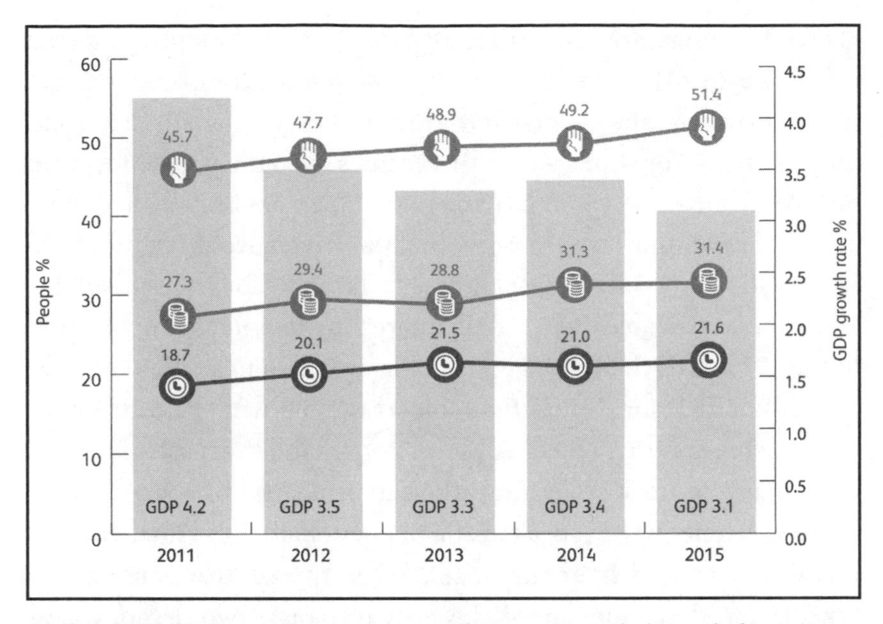

2016 World Giving Index estimates of increases in helping behaviors reported around the world: assisting needy strangers (top line), donating money (middle line), and volunteering (bottom line). The gray bars denote annual GDP growth rate. *World Giving Index 2016. ©Charities Aid Foundation.*

frequently by large proportions of the populace—large pluralities and even majorities in some instances, not just an elite few—is staggering. It's overwhelming.

And not only are people good—they're getting better. Across a wide range of time frames and definitions of altruism, the incidence of helping is continually rising. The World Giving Index shows that trends in all three forms of generosity it evaluates—donating to charity, volunteering, and helping needy strangers—are increasing year to year around the world, although official numbers go back only five years.

Other figures back these numbers up. Charitable giving estimates for the last forty years in the United States show that charitable giving has increased steadily and significantly during this period. Per capita, Americans donated over *three times* as much money to charity in 2015 as they did in 1975, even after adjusting for inflation. Globally, blood donations are also increasing: 10.7 million more donations were made in 2013 from voluntary, unpaid donors compared to 2008. Blood donation rates in the United States also continue to rise, as do bone marrow donations—over three times as many people received marrow from strangers in 2015 as in 1995. It's anybody's guess how current blood and bone marrow donation rates would compare to rates in the more distant past had the technology for widespread donation been available then, but it's interesting to note that it was only in the 1970s that the United States even switched to an all-volunteer blood supply. Prior to that time, blood donors were paid. In other words, whereas 100 percent of blood donors today are altruistic donors, many or most were not fifty years ago. And of course, as recently as twenty years ago, altruistic organ donation did not exist and was widely viewed as unfathomable, even though it was medically feasible and desperately needed. Only in the last two decades have most people even been able to conceive of an act of such extraordinary generosity.

One possible exception to the overall trend may be volunteering in the United States. Bureau of Labor Statistics (BLS) estimates show that volunteering has generally either held steady or dropped slightly during the last twelve years. It's hard to be certain how to interpret this pattern. It is possible that changes in volunteering reflect true reductions in the desire to expend time and effort helping others. Alternatively, because this is the only indicator of altruism that appears to be declining, the trend may instead reflect forces that affect volunteering specifically. It may, for example, reflect increases in the number of hours Americans spend working, which could cut into the time available to volunteer. Or it could reflect the general decline in various kinds of civic participation in the United States, from voting to joining clubs to attending church. Ongoing declines in religious affiliation may be particularly relevant, as religious organizations are the single biggest supporters of volunteering in America. It may be primarily formal volunteering through charitable organizations that is declining, rather than volunteering overall. This would be consistent with the fact that the World Giving Index, which uses a looser

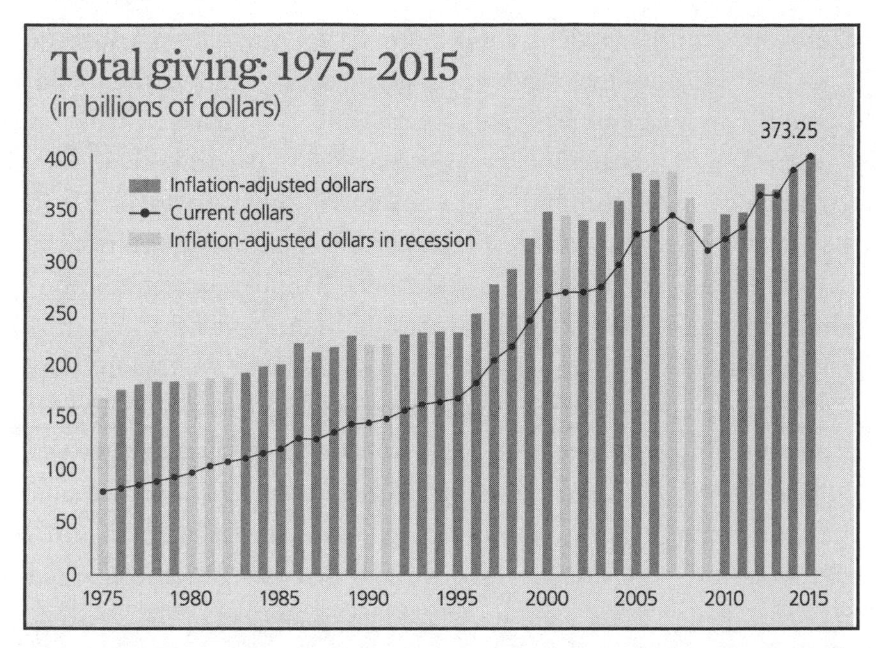

National Philanthropic Trust estimates of annual charitable donations in the United States from 1975 to 2015.

Giving USA Foundation, "Charitable Giving Statistics," in Giving USA 2016, available at: National Philanthropic Trust, https://www.nptrust.org/philanthropic-resources/charitable-giving-statistics/.

definition of volunteering, continues to record increasing, not decreasing, rates of volunteering in the United States.

Not only are people helping one another more, but they are also hurting one another less. In his terrific book *The Better Angels of Our Natures,* Steven Pinker has provided convincing evidence that the incidence of all manner of cruelty and violence has been steadily decreasing for centuries, regardless of the time frame or the type of cruelty under consideration: deaths in international wars, deaths in civil wars, murders, executions, child abuse, animal abuse, domestic violence—the list goes on and on. All of it falling, falling—not linearly, but persistently over time, all over the world. In Europe, the homicide rate today is a mere one-fiftieth of what it was during medieval times. Cruel practices like slavery and torture that were commonplace around the world for many thousands of years are now nearly extinct. Mauritania's abolition of slavery in 1980 marked the first time in history that slavery was illegal everywhere in the world. Torture of even vicious criminals is now widely condemned, whereas it was once a standard punishment everywhere for crimes that today would be considered minor. I recently was stunned to learn that in the United States it was widespread and accepted for police officers to torture suspects up through the 1930s. Support for capital punishment has also been steadily dropping. In 2015, for the first time in fifty years, only a minority of polled Americans still favored the practice; this was also the first year in recorded history in which Europe was completely free of capital punishment. Although 2016 did not see this record matched (Belarus executed at least one person), that year had the distinction of being—once the Colombians ultimately agreed on a FARC peace treaty—the first year in recorded history in which there were no war zones anywhere in the Western Hemisphere.

All of this makes it very difficult to refute that, relative to any reasonable frame of reference, modern human societies are generous, peaceful, compassionate, and continually improving. We could only be considered selfish and violent in comparison to a utopian society in which *no* violence or cruelty takes place—a somewhat unfair comparison considering that there is no evidence that such a society has ever existed. A fairer comparison would be to all the actual alternative realities represented by the various forms that human societies have taken over the millennia. And relative to *any of these actual alternative realities,* the present era is one of overwhelming caring and kindness.

Not that you would ever know this by asking people. Despite all the numbers to the contrary, majorities of respondents polled in the United States and elsewhere believe that people are, as a rule, selfish, preoccupied by their own interests, and untrustworthy—and getting worse. Every year over the last decade a majority of Americans have reported that crime that year increased relative to the previous year, when the precise opposite has nearly always been true. This imaginary dystopia is not just an American phenomenon. Although youth violence and delinquency have also plummeted in the United Kingdom during the last twenty years, majorities of Britons consistently believe that they have increased or stayed the same.

These striking discrepancies between actual reality and beliefs about reality reflect the fact that our brains are not very good at tabulating the actual state of the world. They weren't built to be. Sure, our brains need to calculate the nature of the world around us accurately enough so that we don't walk into walls or fall off precipices. But even our perceptions of the simplest physical surfaces we see in the world around us—a solid white wall, rough ground, a sharp edge—are more illusion than reality. The world as it really exists is a colorless, swirling soup of atomic particles made up nearly entirely of empty space. The rich, textured colors and shapes and feelings that we experience—like solid and white and rough and sharp—*feel* real, but they are not. They are a product of the interpretive machinery inside our brain. Eighty percent of the fibers entering the visual processing areas of the brain emanate from the rest of the brain, not from our eyes. That means the world we see is a warped interpretation of reality, not "reality." And this inaccuracy bleeds into every facet of cognition. Inaccurate perceptions of the world lead to inaccurate memories of it, which lead to inaccurate predictions about the future.

And if our brains can so massively distort our perceptions of simple concrete objects like walls and edges, how badly do you think they distort abstract concepts like "human nature"? The answer is: quite badly.

It would be bad enough if our brains were merely randomly inaccurate in how they perceive, remember, and predict the world, but it's even worse than that. They are also systematically biased toward perceiving, remembering, and predicting bad things. The reason is that, again, brains' reason for existing isn't accuracy—it's survival. As a result, they are especially biased toward focusing on bad things that could threaten our survival over good things that would at best incrementally improve it, a phenomenon known as the *negativity bias*.

The negativity bias dictates that we generally pay more attention to bad events, encode their details with higher fidelity, and remember them better afterward. This asymmetry is as prevalent in the social domain as it is everywhere else. Negative comments from others have a stronger impact than positive comments, such that, for example, the relationship psychologist John Gottman has estimated that a romantic relationship must be marked by at least a five-to-one ratio of kind to unkind comments to be successful. Negative actions also stick in a way that positive ones do not; the worse the action, the more likely it is to be remembered and used to estimate what a person or group of people are really like, with particularly extreme negative actions, like overt cruelty, carrying disproportionate weight. Paradoxically, negative actions carry disproportionate weight in part *because* they are rare and unexpected, which makes them even more attention-getting and memorable when they do occur. As a result, even in a world full of people speaking and acting in overwhelmingly good ways—which they do—we notice and remember the small number of highly callous, selfish, and untrustworthy acts better, and we perceive them as far more representative of reality than is actually the case.

This problem is exacerbated by the fact that much of what we know about the wider world of "people" beyond the ones we know personally derives not from our own experiences, or even from secondhand reports from friends and family, but from the media. This is a problem because the media are also not motivated to represent the actual state of the world in an unbiased way. I'm not referring here to political bias, but to simple negativity bias. Most media outlets are ultimately profit-driven and need people to pay attention to them to sell copies and airtime and advertisements. Because of this, and because people are biased to pay more attention to bad things, the journalists who want us to read or watch or listen to their stories are biased toward telling us about bad things. Bad news sells, as the familiar saying goes. And so, by some estimates, the ratio of negative to positive events covered by popular news media is seventeen-to-one, a ratio that does not remotely reflect the ratio of positive to negative events in the actual world. And it's not just any bad news that sells. As the aphorism goes, "dog bites man" is much less newsworthy than "man bites dog." The more unusual or unexpected the bad news, the more likely it is to capture people's interest and make it to press. Thus, once again, it is in part *because* cruelty and violence are becoming rarer that their newsworthiness continues to increase.

The resulting deluge of selectively reported news about objectively rare violence and cruelty feeds into the perception that we live in a world in which many more bad things happen than good, further fueling the mistaken but common belief that the world is dangerous and becoming ever more so and that people are cruel and callous and getting worse. Is it any wonder that people who consume more news media also tend to be unhappier, more anxious, and more cynical?

An example of this paradoxical process can be seen in the media spotlight that has recently been shining on certain crimes in the United States. For example, sexual assaults on college campuses receive vastly more media attention today than they did in the past. A Google Trends search finds over ten times as many news stories on campus sexual assault in 2016 than there were five years prior. The word "epidemic" frequently crops up in these articles, which probably contributes to the fact that four in ten Americans believe that the United States currently fosters a "rape culture" in which sexual violence is the norm; only three in ten disagree. But this is not remotely true. Sexual assault is no more common on college campuses than among same-aged adults who are not in college. And like most other kinds of crimes, rates of sexual assault are *decreasing,* not increasing, both on campuses and off, according to the Bureau of Justice Statistics. There is no epidemic, other than an epidemic of awareness. Now, this epidemic of awareness may be a good thing if it results in an epidemic of concern about these crimes that contributes to their continued decline—and I hope it does. But the media spotlight is not without downsides, one of which is a massively distorted public perception of the frequency of sexual assault, which in turn worsens broader perceptions about gender relations, law enforcement, and human nature itself.

Hardwired cognitive biases exacerbated by biased media coverage help to explain why people's beliefs about human nature are mathematically incompatible with reality. Amazingly, people's beliefs are even incompatible with the knowledge that they have about *themselves,* although you might assume that self-knowledge would be more resistant to distortion. (It's not.)

In the illuminating Common Cause UK Values Survey, pollsters found, consistent with the results of other surveys, that negativity bias had successfully distorted its respondents' views of human nature. About half of the respondents reported that, in general, people place more importance on selfish values like dominance over others, influence, and wealth than on compassionate values like social justice, helpfulness, and honesty. At least, respondents believed these things to be true about *other* people. When asked about their *own* values, a substantial majority (74 percent) of these same respondents reported that they themselves placed more value on compassionate values than selfish values.

Obviously, one of these two findings has to be wrong. It can't be simultaneously true that most people value compassion more and that most people value selfishness more. So what should we believe: what people say about themselves, or what they say about others? The pollsters took several steps to reduce the odds that the discrepancy resulted from respondents merely bragging or puffing themselves up. They concluded that the primary source of the problem was respondents' overly negative perceptions of others. Ultimately, after comparing people's actual reported values with their perceptions of others' values, the pollsters concluded that a whopping 77 percent of the sample underestimated how much their fellow Brits valued compassion.

Now, 77 percent is high, but it's not 100 percent. Not *all* respondents had equally cynical views of human nature, and the likelihood that a given respondent would underestimate others' compassion was not randomly distributed. A reliable predictor of cynicism about others' values was the respondent's own values: respondents who themselves valued compassion very little also perceived others as valuing compassion very little. Conversely, those who valued compassion more tended to believe that others valued compassion more as well. Psychologists call this pattern the *false consensus effect,* according to which people believe their own values and beliefs more closely reflect what the average person values and believes than is actually the case. As a result, people who are themselves highly selfish tend to believe that others are too, whereas those who are highly compassionate believe that others are as well. Think, for example, of Anne Frank, who concluded, despite all she had seen and experienced, that "people are truly good at heart," or Nelson Mandela, who believed that "our human compassion binds us one to the other." Or Martin Luther King Jr., who in his Nobel Prize acceptance speech said, "I refuse to accept the view that mankind is so tragically bound to the starless midnight of racism and war that the bright daybreak of peace and brotherhood can never become a reality. . . . I believe that unarmed truth and unconditional love will have the final word in reality." Or Mahatma Gandhi, who proclaimed, "Man's nature is not essentially evil. Brute nature has been known to yield to the influence of love. You must never despair of human nature."

None of these people can possibly be accused of naïveté. All knew horrors beyond what any human being should have to witness or experience. But all were people whose compassion for others persisted despite their own experiences, and whose faith in the compassion of others remained undimmed.

By contrast, those who themselves are callous or cruel tend to falsely believe that their values represent the consensus. Compare the beliefs of Frank, Mandela, King, and Gandhi to the beliefs of Richard Ramirez, the notorious serial murderer—nicknamed "The Nightstalker"—who brutalized and killed thirteen people during the 1980s. Although his actions placed him far outside the bounds of normal human behavior, he viewed himself as relatively typical, once asking, "We are all evil in some form or another, are we not?" and claiming that "most humans have within them the capacity to commit murder." The serial murderer Ted Bundy concurred, warning, "We serial killers are your sons, we are your husbands, we are everywhere." Even Adolf Hitler framed his own horrific misdeeds as reflecting basic human nature, retorting, when questioned about his brutal treatment of Jews, "I don't see why man should not be just as cruel as nature." Perhaps Josef Stalin best demonstrated the relationship between the possession and perception of human vice: he once proclaimed that he trusted no one, not even himself.

I realize that many people will persist in believing that people are fundamentally and uniformly selfish and callous by nature, regardless of the objective evidence to the contrary. But the evidence also suggests that rigid adherence to this belief says much more about the person who espouses it than it does about people in general.

Resist the temptation, then, to believe only the most pessimistic messages about human nature. Consider the evidence I've given you, as well as the evidence of your own eyes. Next time you see or read or hear about a callous or terrible thing that some person or group of people has done—or hear someone bemoaning how awful people or just some group of people are—don't succumb to negativity bias without a fight. Stop a moment to remember how genuinely variable people are and ask: Is that terrible thing really representative of people as a whole? Is it likely to even be representative of what *that person* or group of people is like? In some cases it may be—for example, when a psychopath commits a truly heinous crime. But such an act is the diminishingly rare exception, not the rule.

Don't limit your stopping and thinking to the bad things either. The many acts of kindness and generosity that happen every day around all of us can fade into the scenery if we let them. When you see or hear or read about (or commit!) an act of genuine kindness or generosity, please take a moment to notice it and to remember how much goodness there is in the world.

There are many reasons that this approach is worthwhile; perhaps the most important is that trust in others can become a self-fulfilling prophecy, a fact that has been demonstrated using simulated social interactions. Perhaps the most famous such simulation is the Prisoner's Dilemma. In this paradigm, a player is told that he and his partner each have two options in every round of the game. They can choose to cooperate with each other, in which case both will get a medium-sized reward of, say, $3. Alternatively, they can choose to defect. If both players defect, they both get only $1. Where things get interesting is if one player decides to cooperate and the other defects. In this case, the cooperator gets nothing at all and the defector gets $5. The hitch is that the players are not allowed to communicate while making their decisions. They must choose what to do—to trust or mistrust each other—before learning of their partner's decision.

In any given round of the game, the payoff structure ensures that it is always more rational to defect than to cooperate. If his partner defects, a player will get $1 if he also defects, but $0 if he cooperates. If his partner cooperates, the player gets $5 if he defects, but $3 if he also cooperates. And yet, when people play this game, they overwhelmingly cooperate. Why would this be? It happens because the game typically involves multiple rounds—often an indeterminate number of them—and so each player's partner has opportunity to pay him back, for better or worse, as the game goes on. This makes the Prisoner's Dilemma a good model of reciprocity-based altruism. Cooperating in any given round requires a player to make short-term sacrifices that benefit his partner under the assumption that the partner will reciprocate in the future. And in the Prisoner's Dilemma, as in real life, they usually do.

Early studies found that the optimal strategy in the Prisoner's Dilemma is called tit-for-tat: starting out cooperating, then doing whatever your partner did in the last round. If he cooperated, so do you; if he defected, defect right back. Those who use this strategy tend to win out in the long term. The fact that tit-for-tat entails cooperating on the opening move is key. This demonstrates that starting from the assumption that even perfect strangers are probably trustworthy is the more advantageous approach for everyone. Starting from an assumption of others' trustworthiness usually leads to an upward spiral of cooperation and increasing trust.

Trust, in other words, becomes a self-fulfilling prophecy.

In my interactions with altruistic kidney donors, I have gotten a peek into the worlds that such a prophecy can create in real life. Like the most compassionate respondents in the Common Cause UK

Values Survey, altruists' own deep-seated compassion and kindness often leads them to assume the best of others, and to be fairly open with and trusting of others, even people they don't know well. As one altruist averred: "I always say: Everybody helps people in some way or another. There are just different ways to do it." Another concurred, saying: "I think generally people are good, and I think people would like to do the right thing." For my research team and myself, it's been one of the most remarkable aspects of working with them—being treated with the trust and warmth of old friends by people we have just met.

I think this view of the world helps explain altruists' decisions to donate as well. When I ask ordinary adults why they wouldn't donate a kidney to a stranger, they often cite concern that the recipient might not truly deserve it—that person could be a criminal or a drug abuser or otherwise just not quite *trustworthy*. But altruistic kidney donors don't seem to adopt this viewpoint. As one of them told us, "Everyone is going to live their own life and make their own decisions, and some of them are going to be bad and some of them are going to be good. But *nobody* is that bad to not deserve a normal life." Or as another said, "Everyone's life is equally valuable. There's no reason to pick or choose." The fact that altruists are willing to give a kidney to literally anyone means that they must start from the belief that whoever is selected to receive their kidney, it will be someone who deserves life and health and compassion.

You might be tempted to conclude that maybe altruists are just suckers, but that's not it. In one computer simulation study we conducted, altruists were as willing as anyone else to penalize people who actually acted unfairly. But their *default* assumption—their starting point with people who are totally unknown to them—seems to be trust. This approach to the world and the people who populate it seems to result in more positive interactions than would result from a mistrustful approach, and over time they reinforce altruists' perceptions of the basic goodness of the people around them.

Who wouldn't want to live in a world like that?

2. Caring requires more than just compassion.

Understanding that care requires more than compassion is a really, really important aspect of understanding altruism. It suggests that a heightened capacity for compassion is not the only thing that fosters extraordinary altruism. What makes acts of extraordinary altruism—from altruistic kidney donations to my roadside rescue to Lenny Skutnik's dive into the Potomac—extraordinary is that *they are undertaken to help a stranger*. Most of us would make sacrifices for close friends and family members—people whom we love and trust and with whom we have long-standing relationships—but these sorts of sacrifices can be easily accommodated by established theories like kin selection and reciprocity, which dictate that altruism is preferentially shown toward relatives and socially close others and that these forms of altruism are at least in part self-serving. When people violate this dictum by sacrificing for an anonymous stranger, however, their actions suggest that they possess the somewhat unusual belief that *anyone* is just as deserving of compassion and sacrifice as a close family member or friend would be. Think of it as alloparenting on overdrive.

Recent data we collected in my laboratory allowed us to mathematically model this feature of extraordinary altruism. The paradigm we used is called the *social discounting* task, which was originally developed by the psychologists Howard Rachlin and Bryan Jones. Rachlin and Jones were seeking to understand how people's willingness to sacrifice for others changes as the relationship between them becomes more distant. In the task they created, respondents make a series of choices about sacrificing resources for other people. Each choice presents the respondent with two options. They can either choose to receive some amount of resources (say, $125) for themselves or split an equal or larger amount (say, $150) evenly with another person, in which case each person receives $75. In this example, choosing to share would result in sacrificing $50 ($125 − [$150 ÷ 2] = $50) to benefit the other person.

The identity of the other person varies throughout the task. In some trials, the respondent is asked to imagine sharing the resources with the person who is closest to them in their life, whoever that may be. Imagine the person closest to you in your life. Would you accept $75 instead of $125 so that this person could get $75 as well? You probably would—me too. In other trials, respondents are asked to imagine that the other person is someone more distant: their second- or fifth- or tenth-closest relationships, all the way out to their one-hundredth-closest relationship. Typically, the one-hundredth-closest person on anyone's list is not remotely close at all and may be only barely familiar—perhaps a cashier at a local store, or someone seen in passing in the office or church. Now, would you settle for $75 instead of $125 so that someone this distant from you—someone whose name you might not even know—could receive $75? Maybe, maybe not.

Rachlin and Jones, and others as well, find that the choices people make during this task describe a very reliable hyperbolic decline as a function of social distance. This means that people reliably sacrifice significant resources for very close others, but their willingness to sacrifice drops off sharply thereafter. For example, most respondents, when given the choice to receive $155 for themselves or to share $150 with their closest loved one, choose to share. In other words, they will forgo getting an extra $80 for themselves so that their loved one can get $75 instead. This choice indicates that respondents place *even more value* on the sacrificed money when it is shared with a loved one—who otherwise would get nothing—than they would if they had kept it for themselves. But as the relationship in question moves from a respondent's closest relationship to their second-closest relationship to someone in position 10 or 20, the average person's willingness to sacrifice declines by about half. By positions 50 through 100, most people will sacrifice only about $10 to bequeath $75 on a very distant other. This pattern holds up across multiple studies and subject populations and across disparate cultures. It also holds up whether the money in question is real or hypothetical. Rachlin and Jones's term for this hyperbolic drop-off, *social discounting*, refers to the fact that people discount the value of a shared resource as the person with whom it is shared becomes more socially distant.

Can social discounting help to explain the difference between extraordinary altruists, who really do make enormous sacrifices for very distant others, and everyone else? Obviously, money is not a kidney. Sharing it does not require undergoing general anesthesia or major surgery. But in other ways the task is not a bad parallel to donating a kidney. When a living donor sacrifices their own kidney for another person, their choice to give their extra kidney away means that they place even more value on it when it is shared with another person—who otherwise would have no functioning kidney at all—than they do on keeping it for themselves. Think back to kidney donor Harold Mintz's question: if your mother was going to die of renal failure tomorrow and your kidney could save her, would you give it to her? If you answered yes, we can say that you would rather sacrifice half of your total renal resources than leave your mother with none. This is exactly the choice that thousands of living donors make every year. Now, what if the person who needs a kidney is your friend, or your boss, or a neighbor? Would you sacrifice half your renal resources so that they could have some rather than none? If this was a harder choice, you have just discounted the value of your shared kidney.

Our data suggest that social discounting helps us understand the real-life choices that altruistic kidney donors make. The kidney donors and controls in our study—who were matched on every variable we could think of, including gender, age, race, average income, education, IQ, even handedness—completed a version of Jones and Rachlin's social discounting task. Over and over again they made choices about whether they would prefer to keep resources for themselves or share them with close and distant others. Tabulating the results, my student Kruti Vekaria and I first looked at how altruists responded when choosing to sacrifice for the people closest to them. We found that they looked almost exactly like our controls. The data for the two groups overlapped completely, with nearly everyone willing to sacrifice the maximum amount ($85, in this case) to share money with their loved one.

But as we plotted further and further out on the social distance axis, the two groups began to diverge. By their fifth-closest relationship, controls were willing to sacrifice only $65. But altruists hadn't budged. They responded just as they had for their closest loved ones. By position 20, controls' willingness to sacrifice had dropped by about half, to $45, following closely the arc predicted by Jones and Rachlin. But altruists' discounting slope remained so shallow that they were choosing to sacrifice as much for their twentieth-closest relationship as controls were for their fifth-closest. And on and on it went, until the most distant relationship (one-hundredth-closest), by which point controls would sacrifice only about $23—roughly one-quarter as much as they would for a loved one. By contrast, altruists elected to sacrifice more than twice as much as controls—$46—to share $75 with a near-stranger. Their generosity had dropped by less than half.

These results suggest a simple reason why altruistic kidney donors find the decision to share such a precious resource—their own internal organ—with a stranger so intuitive: they don't discount the welfare of strangers and near-strangers as much as the rest of us do. To them, it is nearly as worthwhile to make a sacrifice for someone whose name they don't know or whom they have never even met as it would be for most of us to sacrifice for our closest friends and family. In the words of one donor we have worked with, "I see the world as one whole. If I do something for someone I loved, or for a friend . . . why would I not do it for someone I did not know?" This tendency really does seem to be alloparenting on overdrive, particularly because this generosity emerges even in the absence of vulnerability or distress cues—subjects in the social discounting study never saw or heard another actual person, but only imagined them.

These findings also reinforce the critical distinction between "can" and "does." Altruism is not simply a matter of having the ability to experience compassion and provide care. Nearly everyone *can* be compassionate and caring—at least for some people. The real question is, what do you *do* with that capacity when the person in need of your compassion and generosity is a stranger?

This, of course, leads to another question: can the rest of us flatten out our discounting curves more? Can we become more like extraordinary altruists?

On one level, the answer is almost certainly yes. All the social changes that are already occurring prove it. If people are becoming less violent and more altruistic toward strangers all over the world, then we must all be coming to care more about strangers' welfare than we used to. It would be impossible for any kind of widespread genetically rooted change in the capacity for compassion to have occurred during this period of time, so these changes must reflect cultural shifts instead. Somehow these shifts are causing us to place increasingly more value on the welfare of strangers and flatten our discounting curves—or, as the philosopher Peter Singer and others have put it, to expand our "circles of compassion."

I think of discounting as a mountain on which the self stands at the pinnacle. The slopes of the mountain represent social discounting. If the mountain's slopes are steep, like the Matterhorn, the person at the pinnacle values their own welfare high above that of others, and the welfare of their close friends and family high above the welfare of anyone more distant. Very distant others' needs and interests are down in the foothills and can barely be seen through the haze. What factors might help to compress this mountain a little—flattening its slopes to more closely resemble the gentle silhouette of Mount Fuji—such that the welfare of more distant others is not so steeply discounted?

3. More self-control is not the answer.

Steven Pinker has suggested several possible reasons for ongoing declines in cruelty and violence over time. Some of them might influence social discounting, but others probably don't. For example, one factor that may have contributed to declining violence—but not because it makes us fundamentally care more about strangers—is the rise of centralized governments. Centralized governments oversee the resolution of conflicts and the distribution of resources among individuals—and more importantly, among clans and tribes and nations. When a relatively impartial state mediates disputes, it interrupts the cycles of vengeance and retaliation that erupt when disputes must be resolved by the individuals involved in them. Later, during the Middle Ages, the severe punishments that the state meted out to criminals further reduced the appeal of criminal violence, while the rise of state-regulated trade and commerce increased the appeal of cooperation. According to Pinker, these changes reduced violence for two reasons. First, they shifted the incentives surrounding both cruel and cooperative behavior, rendering violent solutions to provocation or frustration both less necessary and less likely to yield desirable outcomes. Second, these changes may also have changed the social norms surrounding violence. As more wealth and status began to accrue to people who inhibited their aggressive impulses, the ability to exert self-control came to be viewed as more desirable.

Although changes in people's ability or tendency to exert self-control may partly explain declining violence, it almost certainly cannot explain increasing care and altruism toward distant others, because altruism in response to others' distress or need is fundamentally emotional, not rational. As is true for the most common form of aggression—the hot, reactive, and frustrated kind—altruistic urges emerge from deep, primitive emotional structures in the brain. This is clearly true of compassion-based altruism, but it is also true of much learned altruism, and probably of kin-based altruism as well. (Reciprocal altruism is the closest to being genuinely rational, although it too is supported by activity in a subcortical agglomeration, in this case the striatum, that drives reward-seeking.) Ancient subcortical brain structures respond quickly and intuitively to altruism-relevant social cues, like vulnerability and distress in the case of compassion-based altruism. This is probably why altruistic kidney donors overwhelmingly report that their decision to act bubbled up quickly, and in many cases unexpectedly, in response to learning about someone suffering or in need. As one altruist told us, when he first spotted a billboard about someone seeking a kidney, "It was just like I was compelled to do it. The only thing I can figure is God reached down, poked me in the side, and said, 'Hey, go help your fellow man.' . . . It was just overwhelming, I just wanted to do it. I have no clue why." Another decided to donate after wandering by a booth at a health fair and learning about the dire need for kidneys. She recalled thinking simply, *I'm pretty healthy. I have two kidneys. You got anybody that needs one?* Both Lenny Skutnik and Cory Booker recounted that their decision to act was a fast and impulsive response to another person's distress. My own heroic rescuer's decision must have been nearly instantaneous as well—he would only have had a second or so to decide whether to stop and help me. When altruism arises this way, from primitive, emotional processes, the only effect that self-control could possibly have is to suppress it, much as it suppresses aggression.

My colleague David Rand, a behavioral scientist at Yale University, has collected systematic data supporting the idea that generosity toward strangers results from fast and intuitive processes and that rational deliberation suppresses it. He and his students have amassed a wealth of data from experimental simulations, including the Prisoner's Dilemma, showing that people who respond the most generously usually do so quickly and without a lot of thought. The more time people take to stop and reflect, the less generous or altruistic they will ultimately be.

Rand and his colleague Ziv Epstein have also examined the cases of dozens of real-life altruists who were awarded the Carnegie Hero Medal for confronting extraordinary risks to save another person's life. (Lenny Skutnik is one of them.) They wanted to know whether, when facing real risks, people still leap into action first and only later stop to consider the risks, or do they exert self-control to overcome their fear for their own safety? To answer this question, they combed through news archives to find interviews with people who had received Carnegie Hero Medal awards between 1998 and 2012 and extracted fifty-one heroes' explanations of why they had acted. Among these explanations were statements like the following:

> I didn't feel any pain at the time, I think adrenaline kicked in as my only thought was that I had to get to the women, I never felt any pain. The adrenaline just kicked in. I was trying to get there as fast as I could.

And:

> The minute we realized there was a car on the tracks, and we heard the train whistle, there was really no time to think, to process it. . . . I just reacted.

The researchers then had raters evaluate how fast and intuitive versus deliberative and rational each decision was. They also asked them to estimate, based on the details of each situation, how many seconds each rescuer had to act before it would have been too late. Finally, they ran all of the heroes' descriptions through a software program that coded for certain kinds of language, like words and phrases associated with the exertion of self-control.

I'm sure you can guess the results from the descriptions I've given you. Nearly half of the heroes described themselves as having acted without thinking at all, and their descriptions received the highest possible "fast and intuitive" score. Altogether, 90 percent of these altruists received ratings on the "fast and intuitive" end of the scale rather than the "deliberative and rational" end. This was true even for those rescues that had allowed at least a little wiggle room in terms of time—perhaps a minute or two to contemplate whether to act or not. The researchers ultimately found no relationship between how much time was available and how intuitively the altruists responded, suggesting that intuitive responding was not the inevitable outcome of a fast-moving emergency. The computer algorithm confirmed these findings, showing that heroes' descriptions of their decisions incorporated little language suggesting that they had attempted to exert self-control. Together, these findings reinforce the idea that, rather than being deliberate attempts to be noble, urges to care and cooperate are deeply rooted in parts of the mammalian brain that may drive us to act on others' behalf before we fully understand what we are doing or why.

This fact has given me some pause about a growing movement called effective altruism, which is aimed at encouraging people to restrain their initial altruistic impulses in order to accomplish the greatest objective good. The movement was inspired by the work of the philosopher Peter Singer, and its advocates' explicit aim is to convince people to donate to charity only after conducting comprehensive research into the objective impact that their donation will yield. The problem, in Singer's view, is that we are prone to give to causes that happen to tug at our heartstrings—the GoFundMe campaign we saw on Facebook, our local animal shelter, a charity that collects toys for homeless children in our community—rather than rationally planning out altruistic giving to yield the greatest objective good. Instead of helping the GoFundMe family, the pets, and the homeless children, why not use that same money to buy bed nets for dozens or hundreds of families in Africa, reducing their risk of contracting malaria? Wouldn't this result in objectively better outcomes, and wouldn't that be preferable? (Remember how much more good it does to improve the lives of those who start out the worst off?)

I couldn't possibly disagree with the idea of using charitable donations effectively. But I see two problems with the philosophy. First, I doubt that there is usually a way to determine what constitutes the greatest objective good. Many would agree that saving five children from malaria is more valuable than buying supplies for an animal shelter (although others would not), but is it more valuable than donating to a university to support malaria vaccine research? How about supporting research on diabetes, which affects more people at any given time than malaria? Or spending time that could otherwise have been spent fund-raising to prepare for and recover from a kidney donation to save a single person's life? What if that person was a malaria vaccine researcher? Answering these questions requires making so many guesses and assumptions and subjective value judgments that any attempt to arrive at an answer using sheer rationality would quickly spiral into a vortex of indecision.

"A vortex of indecision" is, by the way, where people actually end up when, following a brain injury, they are forced to use only logic and deliberation to make decisions about the future. Such injuries leave IQ and reasoning ability intact, but they prevent the affected from incorporating emotional information from deep within the brain into their decision-making. It turns out that intellect and reasoning alone are not sufficient for making complex subjective decisions. People who cannot generate an intuitive feeling of *caring* about one outcome more than another struggle for hours to make decisions about even basic things like which day of the week to schedule a doctor's appointment—the kind of decision, like so many others, for which there is no purely rational answer.

As he accepted his Nobel Prize for literature in 1950, the philosopher Bertrand Russell declared, "There is a wholly fallacious theory advanced by some earnest moralists to the effect that it is possible to resist desire in the interests of duty and moral principle. I say this is fallacious, not because no man ever acts from a sense of duty, but because duty has no hold on him unless he desires to be dutiful." Ultimately, the gut-level, irrational feeling of just *caring* more about certain causes than others is what moves people to help. Desire, not reason, drives action. This is why even the most sophisticated computers don't yet act of their own accord, despite having perfect reasoning abilities—they have no feelings or desires. Psychopaths are often highly rational, but this does not drive them to provide costly help to others, believe me, because they lack the emotional urge to do so. And those who do go to great lengths to help others overwhelmingly describe their motivations in terms of impulse and feelings. Consider the case of Robert Mather, the effective altruist and founder of the Against Malaria Foundation, which has been called the most "effective charity" in the world. In Mather's own telling, he was first moved to devote his life to charity work, not by clear-eyed rationality, but because he stumbled upon the story of a single little girl who had been horribly burned in a fire, and whose story moved him to tears.

Even when people do describe their decisions in terms of clear-eyed rationality, their brains may tell a different story. One altruist who participated in our research described his decision to donate in beautifully utilitarian terms. Upon first reading a news article about altruistic donations, he said,

> it clicked with me immediately, and then I thought: this is something that I could do, and something I was comfortable with. So at that point, I did some research on the web about side effects and mortality rates and the possibility of comorbidities afterwards, and I was very comfortable that the risks were low and acceptable to myself, and that the benefits to the patient—especially if they were already on dialysis—the improvements in lifestyle and improvement in life span and their ability to get back in their life, the benefits were great.

He conducted, in other words, a simple cost-benefit analysis. When I asked about any other thoughts or feelings he might have been having that contributed to his decision to donate, he replied, "I guess I would say I am super-rational—I do not get emotional about the decisions I make." On one of the standard empathy scales we used in the study, he reported his own levels of empathy to be very low. I had no reason to doubt anything he said about himself. It was clear from his professional accomplishments in the technology sector and the way he described other decisions he had made that he was capable of sophisticated rational analysis. But our data revealed that rationality was not all that he was capable of.

When we first examined the scatterplots describing our brain imaging and behavioral data, one of the altruists had stood out from the rest. Of the nineteen we had tested, one scored nearly perfectly in terms of his ability to recognize fearful facial expressions—the top scorer of all our participants. This person also showed a robust amygdala response to fearful expressions during the brain scan—easily in the top half of the altruists. Who was this super fear-responder? None other than our self-described super-rational, low-empathy altruist. I believe this altruist considered himself to be low in empathy. And he may well have been relatively low in the sort of cognitive empathy that is linked to Theory of Mind and autism. But he also had remarkably *high* levels of the kind of empathy that is important for caring and altruism: sensitivity to others' displays of vulnerability and distress.

Of course, this one data point cannot prove that this altruist's heightened sensitivity to others' fear was the cause of his extraordinary actions. But it does prove that you should never take people's self-reported empathy at face value. Just as Daniel Batson's subjects were led to believe that a placebo called Millentana had led them to behave altruistically, so can all of our brains easily mislead us about the causes of our own behavior and feelings and decisions.

Everything we know from the laboratory suggests that deliberation and rationality are not what ultimately drive people to care. Indeed, the more deliberatively and rationally people think about generosity, the more likely they may be to suppress their initial urges to help, and the less generous they ultimately become. Viewing people's natural desires to help particular causes as a springboard to action, rather than as something to be suppressed or overridden, seems to me like it would be a more genuinely effective approach than insisting on pure altruistic rationality.

4. Key cultural changes have made us more caring.

So if it's not self-control that is leading to more caring and compassion in the world, what could it be? Another possibility is that a much more general change—one that has also been indirectly promoted by the rise of state governments quelling violence and promoting trade—is responsible: an increase in quality of life. People act better when they are themselves doing better.

The last millennium has been a period of extraordinary improvements in human prosperity, health, and well-being around the world. It's not just deaths and suffering from violence that have decreased during this period—it's deaths and suffering from all causes, including famine, injury, and disease. Global hunger has precipitously declined. Life expectancies have more than doubled over the last 200 years. Near-miraculous advances in medicine have eradicated horrifying diseases, like smallpox, plague, polio, and measles, that once ravaged millions of people around the world. Do you know what scarlet fever is? I don't, and neither do you, probably. But as recently as 150 years ago, epidemics of it killed tens of thousands of children every year—sometimes every child in a family in just a week or two. Two of Darwin's children were killed by it, as was John D. Rockefeller's grandson. It's one of dozens of former scourges that are now all but gone. Only fifty years ago, one child of every five born around the world died before their fifth birthday. The rate is now less than one in twenty-five. It is easy to miss the significance of these changes because they have been so gradual and consistent. But the amount of human suffering and misery that has been alleviated in the last century is, in reality, staggering.

Education rates also continue to improve worldwide. Literacy was near 0 percent essentially everywhere in the world 500 years ago. As recently as 1980, barely half of the world's population could read. But the global literacy rate is now around 85 percent, and across broad swaths of the world it is close to 100 percent, thanks in part to tremendous strides made in public schooling and in providing equal educational opportunities for boys and girls.

Wealth is increasing at astonishing rates as well. The economic historian Joel Mokyr has observed that in modern industrialized nations, middle-class families enjoy higher standards of living than emperors or popes did just a few centuries ago. The unequal distribution of wealth remains a serious concern, but the poor are also better off than they used to be. The proportion of people living in abject poverty continues to fall all over the world, dropping from around 90 percent of the global population in 1820 to just under 10 percent today. The World Bank estimates that in just the three-year span from 2012 to 2015, the number of people living in extreme poverty (defined as living on less than $1.90 per day) dropped by 200 million, bringing the percentage of people living in extreme poverty to below 10 percent of the global population for the first time. That is a remarkable amount of progress in a very short time. World Bank president Jim Yong Kim called it "the best story in the world today."

There is every reason to believe that these increases in prosperity and quality of life have been the source of many other positive downstream effects—which include ongoing positive trends in generosity and altruism toward strangers, up to and including extraordinary altruism. That well-being has increased worldwide in tandem with various forms of generosity and altruism toward strangers is clear, although obviously this is merely a correlation, and a wildly confounded one at that. But my lab and others have conducted more targeted research showing that, even after controlling for many possible confounds, increasing levels of well-being are associated with increased altruism.

A few years ago, my student Kristin Brethel-Haurwitz and I were combing through national statistics on altruistic kidney donation when we noticed the incredible variation in rates of donation across the fifty US states. We wondered why this might be. Around the same time, the Gallup polling organization came out with its first-ever statistics on variations in well-being across the states. When we compared maps of altruistic kidney donations and well-being side by side, the similarities were striking. We ran a number of analyses to probe these similarities and found that even after we controlled for every difference we could possibly think of across the states—median income, health metrics, inequality, education, racial composition, and religiosity, to name a few—high well-being in a state remained a strong predictor of altruistic kidney donations.

Well-being is more than just happiness—it's life satisfaction, having a sense of meaning and purpose, and being able to meet basic needs. These are qualities shared by denizens of well-off states like Utah, Minnesota, and New Hampshire, which, though very different in some ways, all produce high proportions of altruistic kidney donors. In states like Mississippi, Arkansas, and West Virginia, on the other hand, both well-being and altruistic donations are very low. Kristin and I also found that although well-being is somewhat

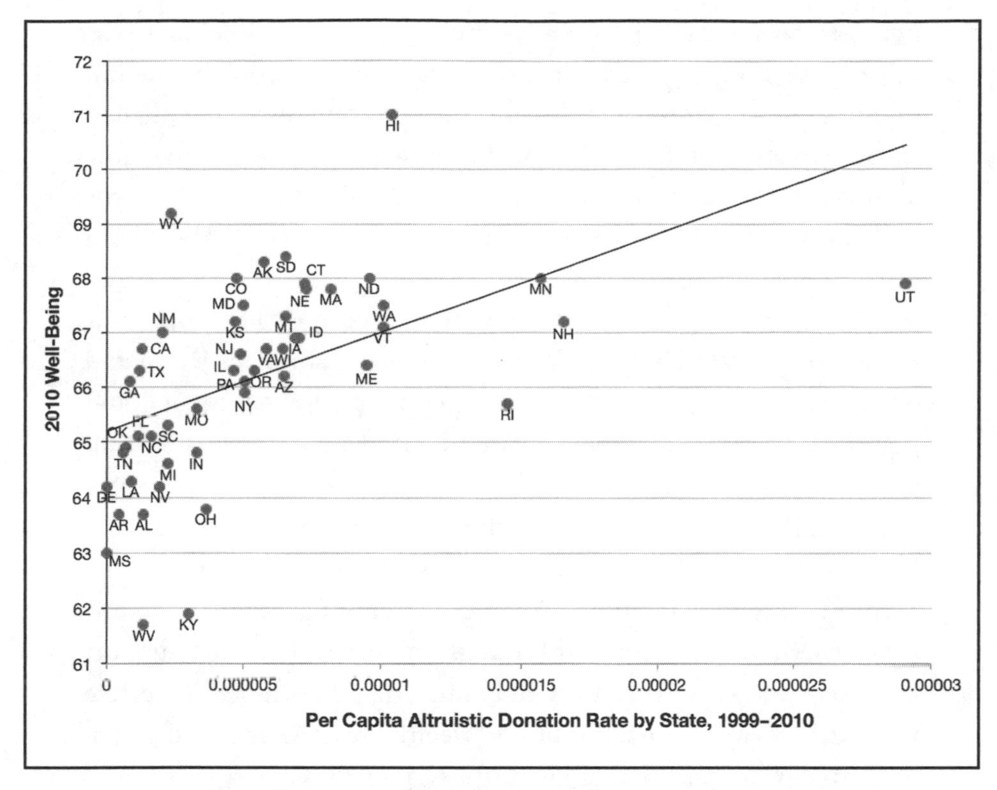

Across the fifty US states, rates of altruistic kidney donation rise in tandem with well-being. *Abigail Marsh and Kristin Brethel-Haurwitz.*

related to baseline variables like income and health, it's even more strongly related to whether these indicators are *improving*. We found that, even after controlling for baseline income and health, increases in median income and health over a ten-year period were strong predictors of both well-being and extraordinary altruism.

In some ways we found this result surprising. It's a common trope —embodied by super-wealthy fictional characters from Ebenezer Scrooge to Gordon Gekko to the Malfoys—that wealth and status lead to selfishness. But these tropes are not really relevant to our findings. "Wealth" in large population studies like the one we conducted doesn't refer to people with butlers and mansions. The super-wealthy represent only a tiny fraction of the population, and their actions aren't reflected in our data set. Instead, our findings suggest that incremental increases in objective and subjective well-being across large groups of people also increase altruism. As people move out of poverty and into the middle class, or inch from below the median income to above it, the odds that they will opt to give a kidney to a stranger fractionally increase. This says more about possible benefits of financial security and reductions in poverty than it does about stereotypical wealthy people.

Our findings are consistent with a large body of literature linking generosity to well-being. The psychologists Elizabeth Dunn and Mike Norton, as well as others, have conducted experimental and population-level studies that consistently find a positive relationship between well-being and various forms of generosity, such that people who report higher well-being, or whose well-being is experimentally tweaked, tend to behave more generously. This is in keeping with the theory that flourishing promotes engagement in a variety of voluntary, beneficent activities. Our findings on kidney donations support this theory, as do a wide array of studies that have linked objective measures of well-being, including wealth, health, and education, to everyday generosity and altruism. One 2005 Gallup poll found a linear relationship between income and volunteering, donating money (of any amount), and donating blood. Individuals in households earning more than $75,000 per year were the most likely to engage in all three behaviors, followed by those living in households earning more than $30,000 per year, then by households earning less than that amount. (Keep in mind, of course, that such studies tell us about population *averages* rather than the behavior of any one individual; plenty of less-well-off households are generous, and plenty of wealthier ones are not.) Another large study found similar results in Canadians: the best predictors of charitable donations, volunteering, and civic participation included higher income and more education. Large-scale naturalistic experiments reveal similar patterns. A field experiment in Ireland found that socioeconomic status was the best predictor of donations to a child welfare charity and of altruism. In a "lost letter" paradigm (first created by Stanley Milgram, incidentally), stamped letters addressed to charitable organizations were left on the ground for passersby who were so inclined to pick up and deposit in a mailbox. As in similar previous studies in the United States and England, letters dropped in more deprived neighborhoods were less likely to be returned than letters dropped in less deprived neighborhoods. The positive relationship between well-being and altruism persists across cultures, from Taiwan to Namibia.

One reason for these patterns may be that lower levels of well-being that follow from financial insecurity, poor health, or traumatic life events inhibit altruism by souring people's view of the world, human nature included. Many decades of research on misanthropy find that this trait, which reflects cynical attitudes about human nature and lack of faith in other people, is inversely related to most indicators of well-being. People who are experiencing hard times are much more likely to report dour views of others, such as that others are only looking out for themselves and cannot be trusted. This suggests that, among the many positive sequellae of being wealthier (again, in the less-poor sense, not in the mansions-and-butlers sense), healthier, and enjoying higher social status appears to be a greater tendency to view others as generally trustworthy, kind, and generous.

To be fair, some recent studies conducted by psychology researchers whom I greatly respect have found conflicting results. For example, a study conducted by the psychologists Paul Piff, Dacher Keltner, and their colleagues found that people who drive luxury cars (who also tend to be wealthier) were less likely than other drivers to follow traffic laws and norms like yielding at a four-way stop or at a pedestrian crossing. In other studies, undergraduate students at the University of California–Berkeley who placed themselves higher on a ladder representing their overall standing in the community were less likely to share resources with strangers in a computer-based economic game. Parallel results were obtained using a national sample of adults from an email list maintained by a private West Coast university; of these adults, the wealthier and more educated ones were less generous in a computerized task. And in a sample of adults recruited from Craigslist, those who reported their social status to be higher were more likely to cheat in an online game of chance.

I myself was quite torn about these divergent sets of findings, being just as familiar with the trope that wealth promotes selfishness as everyone else. Then I encountered the biggest, most ambitious, and best-controlled study conducted on this topic yet, the results of which were *so* clear and *so* consistent that they convinced me that being better off may—again, on average—actually increase a wide variety of caring behaviors toward strangers. The study was conducted by the German psychologist Martin Korndörfer and his colleagues, who were familiar with the thinking that wealth and status tend to promote selfishness. So they conducted eight large studies that aimed to examine the association in large samples. And I mean *really* large: their studies included upwards of 37,000 people. Importantly, these were also representative samples, which capture the behavior of an entire population, not just selective subsets of it. All else being equal, findings from larger and more representative samples are more likely to be accurate than findings from smaller, more selective samples.

The researchers were surprised to find the opposite of what they had expected. They started off looking at charitable donations in their native Germany and found that, as wealth and education and status rise, Germans donate proportionally more of their income to charity, not less, and that the proportion of households that donate increases as well, from around one-quarter of the poorest 10 percent of households to three-quarters of the wealthiest 10 percent. They next looked at Americans' charitable donations and found exactly the same effects. They found similar patterns again for volunteering—wealthier, higher-status Germans and Americans were more likely to volunteer to help others, and they volunteered more frequently. Reinforcing the idea that "wealthier" in these studies is not referring to the super-wealthy, the researchers found that generosity increased consistently with wealth along the entire income spectrum, from the very poor to the slightly less poor to the middle class to the wealthiest, who were still nowhere near super-wealthy. (In the United States, the top 10 percent of households earn around $160,000 or more per year, which is well-off, to be sure, but hardly super-wealthy.)

The researchers also looked at everyday helping behaviors in large, representative US samples—behaviors like carrying a stranger's belongings, or letting them go ahead of you in line, the forms of altruism that are most likely to be spontaneous reactions to another person's need—as well as behavior in a controlled economic game in which resources could be freely given to a stranger. The pattern was always the same. Those who were relatively better off were more likely to give. Finally, consistent with the literature on misanthropy, the economic game also showed that wealthier and higher-status players were not only more trustworthy (giving more resources to the other player) but more trusting as well.

These findings contradict the possibility that wealth and status are correlated with generosity only because poorer, lower-status households have fewer resources to give away. If it were only the case that poor households donate less money than wealthier ones, this explanation would make sense. But the fact that both the proportion of income donated and the likelihood of donating anything at all continue to rise with every incremental increase in wealth and status does not fit as well with this explanation. Surely most middle-class families have at least *some* resources they could donate—but the likelihood that they will donate anything rather than nothing increases at every point along the spectrum of wealth and status. So too does the likelihood that they will volunteer their time, despite the fact that wealthier individuals generally have less leisure time, not more. Moreover, there is no clear reason why poverty would impede everyday helping behaviors like giving directions or helping someone carry their belongings.

Could these patterns be somehow unique to Germany and the United States? It appears not. When the researchers examined patterns of volunteering in twenty-eight other nations across five continents, they found identical results in twenty-two of them (interestingly, exceptions included states with strong social welfare systems, like France, Norway, and Sweden, where volunteering was roughly equal across incomes), and in no country was increased wealth associated with less generosity.

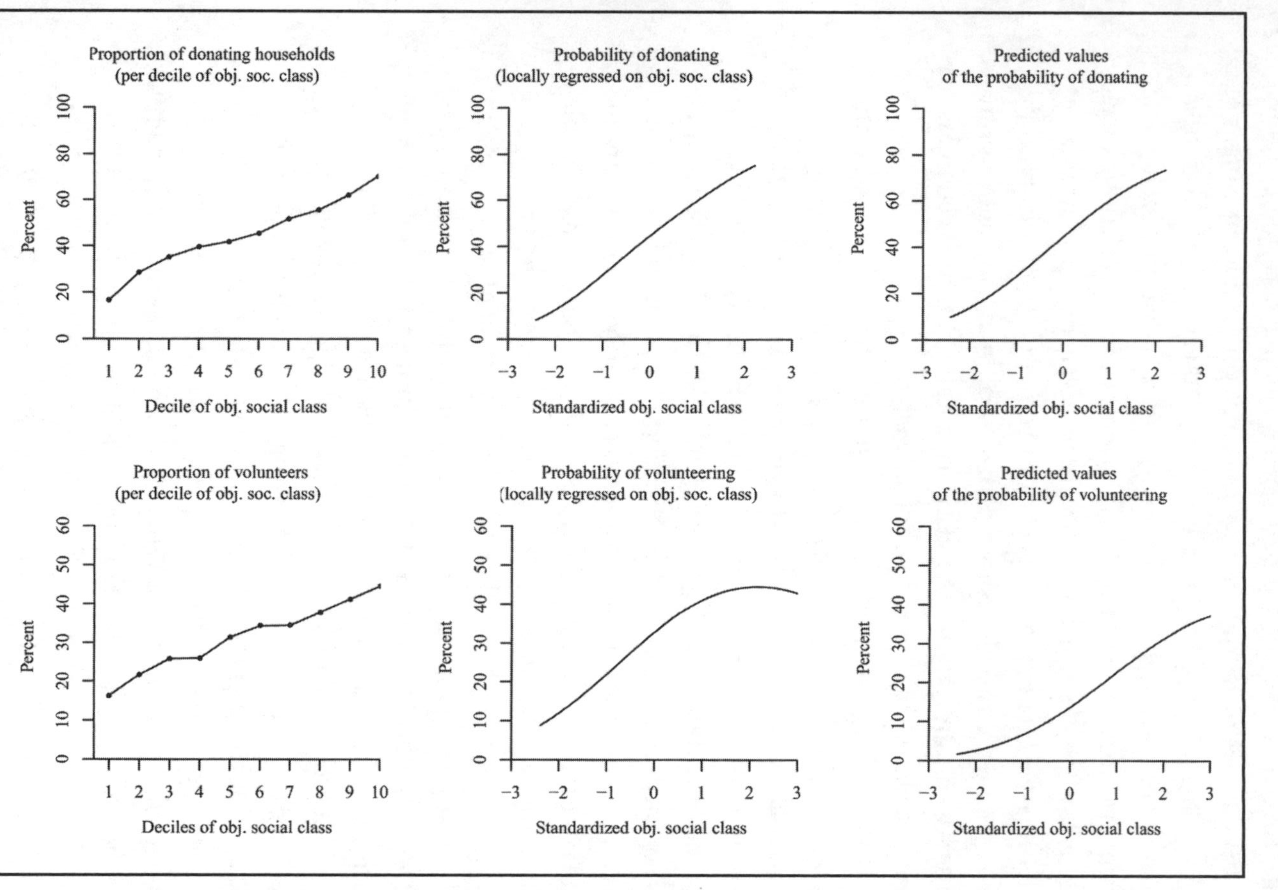

Americans' likelihood of volunteering and making charitable donations as a function of objective social class (calculated based on income, education, and occupational prestige). *M. Korndörfer, B. Egloff, and S. C. Schmukle. "A Large Scale Test of the Effect of Social Class on Prosocial Behavior."* PLoS One *10, no. 7 (2015): e0133193.*

All in all, the data are overwhelmingly clear. Studies that draw from large, unbiased population samples all show the same effects: those who are relatively more prosperous are more likely to donate blood, donate money, volunteer to help others, engage in civic activities, engage in everyday helping behaviors, return lost items to strangers, behave generously in computerized simulations—and even engage in extraordinarily altruistic behaviors like giving strangers their own kidneys. In general, the studies that find otherwise draw from smaller and more selective samples and so may reflect inadvertent biases inherent in these samples, or in the way social status was measured.* In some ways, these findings come as a tremendous relief. The world *is* becoming wealthier, more educated, and more prosperous. These trends show no sign of slowing or stopping. Can you imagine if all of these changes inexorably led to people becoming more and more selfish as well? What a terrible Faustian bargain it would be, to have to choose between people becoming better off and becoming better hearted. But I don't think we do.

There is a big caveat to these findings, however. All the metrics of altruism across the various studies I've described share one thing in common that may not be immediately apparent. They focus on altruism toward *strangers:* donating to charities that help strangers, volunteering to help strangers, mailing letters to strangers, donating blood and kidneys to strangers, and so on. None of these data tell us anything about people's generosity toward their family, friends, and neighbors. So it isn't at all correct to draw the conclusion that people who are better off are more compassionate or generous or altruistic *overall.* There is no evidence that this is true. The evidence shows

* Some previous studies identified a u-shaped curve for donations as a function of income, with very low- and very high-income households contributing the most. The analyses of Körndorfer and his colleagues indicate that this may be an artifact of analyses that exclude nondonor households: those that donate nothing. When both donor and nondonor households are analyzed together, to account for the fact that fewer lower-income households make any charitable donations at all, the u-shaped curve becomes a simple linear increase, with donations consistently increasing with household wealth and status.

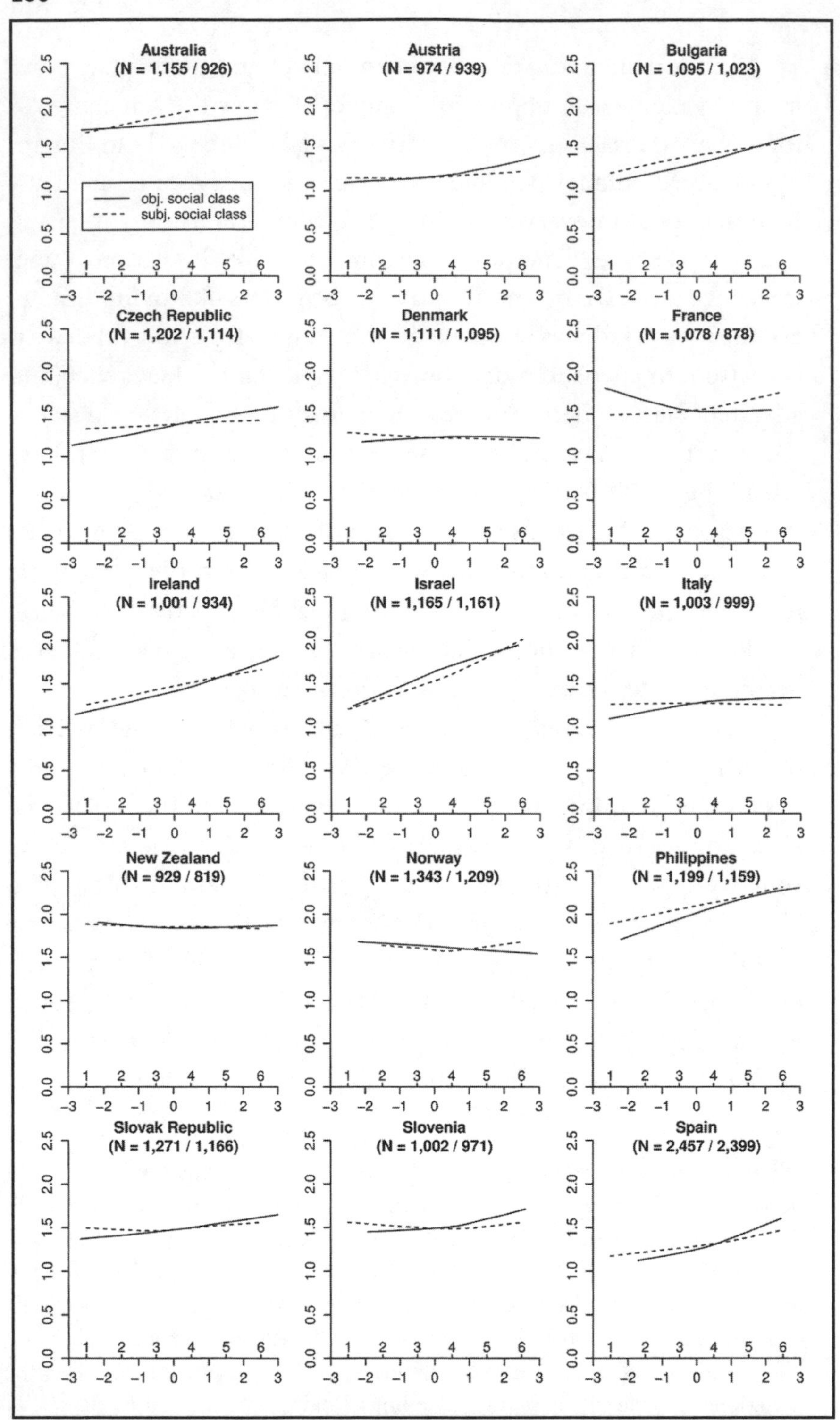

The frequency of volunteering across thirty nations as a function of objective (solid line) and subjective (dashed line) social class.

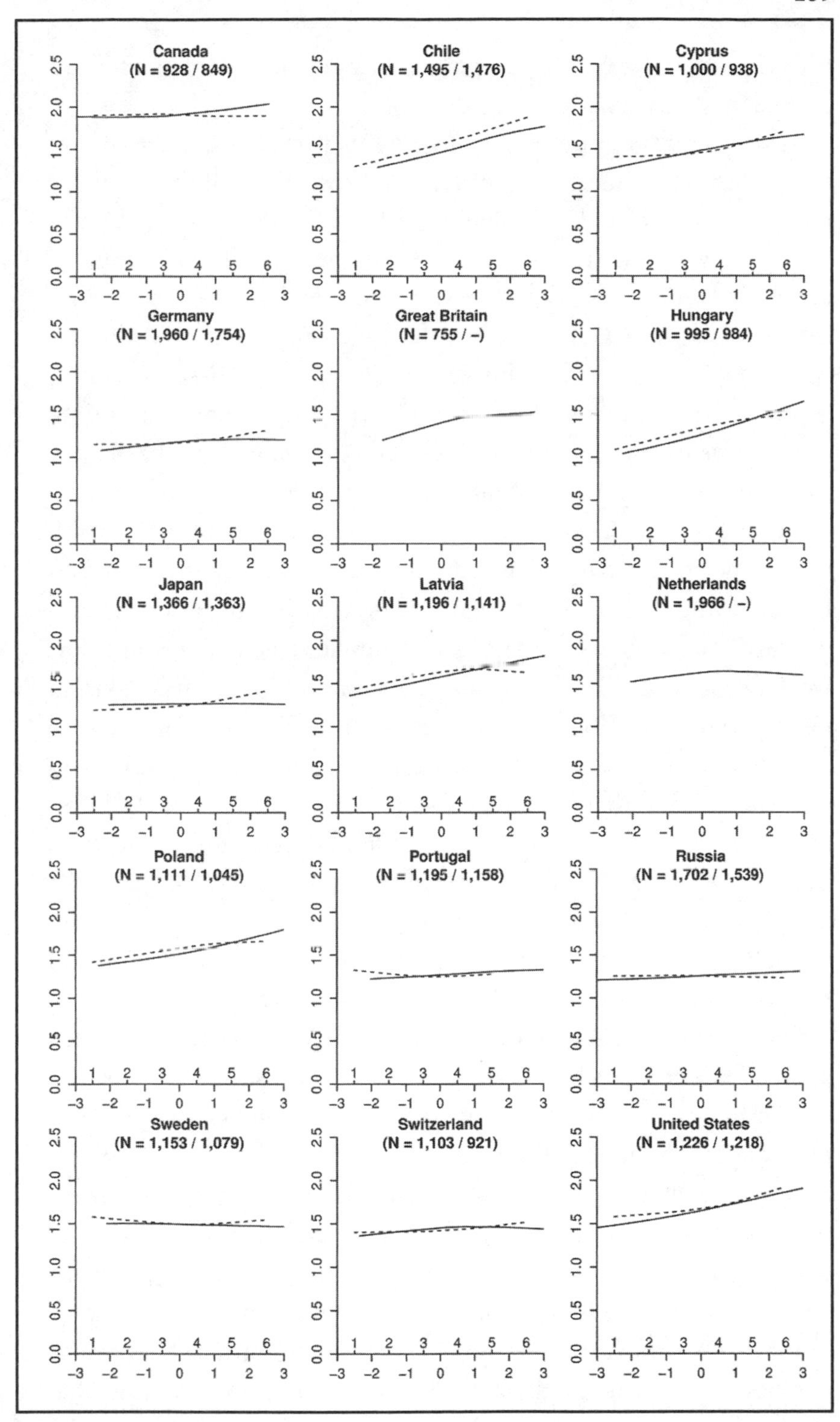

M. Korndörfer, B. Egloff, and S. C. Schmukle. "A Large Scale Test of the Effect of Social Class on Prosocial Behavior." PLoS One *10, no. 7 (2015): e0133193.*

only that as people become better off, they are more altruistic toward people they don't know.

This is an important detail, and one that helps to make sense of the patterns that have been observed. Prosperity within a culture tends to be associated with characteristic cultural values and norms that influence how strangers are viewed and treated, including relatively less emphasis on collectivist values and more emphasis on individualist ones. Collectivism entails focusing on and valuing interdependence within a family or community, a mind-set that is essential when resources are scarce and strong social connections are critical to survival. By contrast, individualist values emphasize the independence of the individual and personal instrumental goals.

Analyses of cultural values across nations reliably find wealth to be a positive predictor of individualism, with wealthy countries like the United States, Australia, the United Kingdom, and the Netherlands having some of the highest individualism scores and less wealthy countries like Guatemala, Ecuador, Indonesia, and Pakistan having some of the lowest. And individualist values tend to increase with growing prosperity within a given culture. For example, one recent study found that over the last 200 years, indicators of individualism increased in tandem with wealth in both the United States and the United Kingdom. And the recent period of incredible economic growth in China, historically one of the world's most collectivist nations, has also coincided with a growing emphasis on individualism there. This effect may be bidirectional, such that wealth and individualism are mutually reinforcing. Prosperity has reliably predicted subsequent increases in individualism over the last 150 years of cultural changes in the United States, but individualist values may also promote economic growth.

It can be tempting to think of individualism as a proxy for selfishness and collectivism as a proxy for generosity, but this is not exactly true. Collectivist cultures do very much value social ties, generosity, and cooperation—but primarily toward members of their own communities. In China, for example, traditional Confucian teachings emphasize the value of altruism, but with a focus on altruism that benefits close family members and friends. And in Fiji, which is also a highly collectivist society, people tend to be extremely generous toward others in their village. But according to the behavioral scientist Joseph Heinrich, when "Fijians do games that involve giving to distant poor people, they seem almost baffled as to why anyone would send money to someone they don't know far away." An emphasis on group bonds requires that members of collectivist cultures draw clear distinctions between the group members whose welfare, goals, and identities are deeply interdependent and everyone else. And relatively little value is placed on the welfare of everyone else. This is an unfortunate and perhaps inevitable* downside of maintaining strong group bonds and identities. Collectivism is associated with low levels of what is called *relational mobility,* meaning that relationship networks in collectivist societies are not only strong and interdependent but also stable across time. A collectivist can assume that their closest relationships will remain their closest relationships for years or decades into the future. There are clear upsides to such stability, but one downside is the way it affects people's attitudes toward anyone new. In individualist cultures, higher relational mobility means that anyone unfamiliar could "one day become a friend," as cultural psychologist Yulia Chentsova-Dutton, a colleague of mine at Georgetown, put it to me. Her words struck me as a beautiful description of my initial interactions with many of the altruists I've met. They seem to approach us right from the start not as strangers but as potential friends. This is an approach that is less common among collectivist cultures, where it is more likely to be assumed that a stranger will stay a stranger.

Decades of social psychology research also make it exceedingly clear that dividing people into clearly defined groups is a great way to get them to treat members of other groups worse. This is true even if the groups are nonsensical—in what are called *minimal group paradigms,* people can be instantly induced to downgrade the value of a

* Not completely inevitable, fortunately. There is a simple and very effective way of reducing hostility between groups: simple intergroup contact. The more we interact with members of perceived outgroups, the more we tend to like them.

stranger's welfare when merely told that the stranger is a member of an arbitrarily created "Blue" group instead of the subject's own "Green" group. And when the wider culture paints members of an outgroup as actively threatening or contemptible—think of how Nazi-era Germans depicted the Jews, or how many modern nations view Muslim refugees—compassion toward members of those groups will be further suppressed. In such cases, the fear *of* others drowns out any inclination to fear *for* those others' welfare. This phenomenon can be exacerbated by a negativity bias–driven media, which is a powerful purveyor of cultural norms and helps determine who is viewed as a threat. Again, however, poor treatment of outgroup members is not a reflection of anyone's *capacity* for compassion. If anything, compassion for our loved ones can sensitize us to anyone we view as a threat to them—much as the lioness chased away her fellow lions when they encroached on her baboon "cub," or as Mimi the chihuahua became hostile toward her owner once she had "pups" to protect.

These psychological phenomena may help to explain why, although members of individualist cultures are not more altruistic overall, they do tend to be more altruistic, on average, toward strangers. Individualist cultures tend to rank higher on the World Giving Index, which, again, measures charitable donations, volunteering for social organizations, and providing everyday help—for strangers. The top of this index is reliably dominated by the world's most individualist nations, including Australia, Canada, Ireland, the Netherlands, New Zealand, the United Kingdom, and the United States. These countries are also vastly more likely to donate blood, bone marrow, and organs—at least deceased organs—to strangers. (Living organ donations to strangers are still too rare to meaningfully compare across cultures, but the vast majority of deceased organ donations are performed in individualist nations in Europe and North America.)

Some of these differences clearly reflect wealthy nations' infrastructure and outreach advantages—it's hard to run a blood bank or a transplant center without a reliable power supply or funds to pay staff or buy supplies—but cultural views also seem to play a role. In studies of blood donations conducted in relatively collectivist countries in Africa and Asia, respondents often say that they would donate blood to a family member, but much smaller fractions would consider donating to a stranger. This may contribute to the persistent problem in many developing nations of low blood donation rates (averaging 0.4 percent—too low to meet a nation's basic blood requirements) and near-total reliance on donations from relatives or for pay. Even within an individualist nation like the United States, members of more individualist subcultures are more likely to volunteer to help strangers or donate organs to them than members of more collectivist subcultures.

Again, none of these broad cultural patterns are anywhere near absolute—they represent average differences across groups that help us understand the cultural forces that cause widespread changes in compassion and altruism. They cannot predict how altruistic any given person will be. Plenty of wealthy people in individualist cultures remain impervious to the needs of strangers, and members of poor or collectivist cultures can be overwhelmingly generous. Exceptions to the overall patterns can result from, among other things, specific religious or cultural views that promote generosity toward strangers. For example, Myanmar is a nation that is neither wealthy nor individualist, but it reliably ranks number one on the World Giving Index, thanks in part to the widespread practice there of Theravada Buddhism, which emphasizes the importance of giving.

Myanmar is also notable for one additional feature that may not be incidental to altruism: it has a relatively high literacy rate (over 90 percent). Literacy is yet another correlate of increased prosperity that may heighten genuine concern for the welfare of distant others.

The current age of widespread literacy was initiated by the invention of the printing press at the dawn of the Renaissance. This invention represented a new era for many reasons, one of which was, of course, that the mass production of books allowed knowledge to be disseminated more easily and cheaply. But that wasn't all. Books are not merely knowledge containment units—they are not manuals. No one would read them if they were (see: manuals). Instead, books are windows into the minds of the people who wrote them and the people who are written about. Fiction, in particular, represents what the psychologist Keith Oatley calls "the mind's flight simulator"—a vehicle for exploring the rich mental and emotional landscapes of people we have never met.

In making the experiences and emotions of people from vastly different and distant cultures both emotionally transporting and relatable, fiction enables us to become emotionally invested in the characters we encounter, to care about their plights and their fates. Although any kind of fictional media—including movies, television, or radio—could theoretically achieve this effect, written fiction may be uniquely powerful for breaking down barriers between cultures and groups. This is in part because a person represented through written words is completely abstracted and uncontaminated by the foreign accents, styles of dress, and mannerisms that can mark physically embodied people as outgroup members whose welfare might be devalued during face-to-face interactions. In allowing readers to perceive the world from inside a disembodied stranger's head, fiction provides people across cultures with a visceral appreciation for the universality of their own emotions and experiences, thereby lowering barriers to compassion.

Steven Pinker has made a strong argument that the emergence of literacy played a major role in historical declines in violence, probably by directly strengthening people's ability to care for distant others. Supporting this possibility, laboratory studies show that exposure to the written word can increase empathy and compassion for strangers. In one study conducted by Daniel Batson, some subjects read a brief note in which a fictitious stranger, whom subjects never personally saw or met, described her distress over a recent breakup: "I've been kind of upset. It's all I think about. My friends all tell me that I'll meet other guys and all I need is for something good to happen to cheer me up. I guess they're right, but so far that hasn't happened." Subjects who read these words and then played a single-trial Prisoner's Dilemma game with the stranger who wrote them showed impressive increases in generosity toward her—*even though she had just defected against them!* Twenty-eight percent of the subjects who read the note chose to cooperate, despite defection in this scenario being the objectively optimal response. Compare this with the number who cooperated with her without having read her written correspondence: 0 percent. The stranger's words made her human—and a vulnerable, distressed human at that.

More recent research has also shown that the empathic effects of reading generalize past the specific individuals depicted in the text. People who read more fiction (but not nonfiction) are better at identifying complex and subtle emotions in others' faces. And when subjects in one study were experimentally assigned to read a work of literary fiction, they reported increased empathic concern for others even long after they had closed the book.

It seems, then, that a potent combination of steady cultural changes over the last several hundred years of human history may be responsible for the decreasing violence and increasing altruism toward strangers during that period. The rise of states suppressed violence and promoted trade, leading to increasing standards of living and abundance of resources. Growing abundance has gradually reduced the extent to which people must rely on strong, closed social groups for survival and permits a loosening of the severe distinctions between social group members and outsiders. Abundance has also promoted increases in education, particularly literacy, which has further encouraged the sharing of the fruits of growing abundance broadly rather than narrowly: the slopes of the mountain of social discounting are flattening.

Together, these changes help to reinforce why, although the human capacity for altruism reflects basic biological processes and is highly heritable, cultural forces can also cause altruism to increase. (We already know the same is true for human height, which is also highly heritable and has also been increasing around the world for decades due to cultural factors, a phenomenon known as the Flynn Effect.) The genes that build the brain structures that motivate us to care are not operating in a vacuum. When vast and rapid shifts in altruistic behavior are observed across time or nations, these changes cannot be due to changes in the genome—only to changes in the culture in which that genome is expressed. The structures that our genes build are under the constant influence of cultural forces that influence how much compassion and caring any given person will exhibit—and for whom.

8

PUTTING ALTRUISM INTO ACTION

A VARIETY OF cultural shifts—prosperity, cultural norms, literacy—represent sea changes that have affected patterns of violence and compassion across large populations over long periods of time. But what about promoting altruism in the shorter term? Are there strategies that can effect meaningful changes at the individual level?

Again, happily, the answer is yes. One piece of evidence for this actually derives from the cultural data. As cultures become more individualist, it's not just the kinds of behaviors that people engage in that tend to shift, but also all the motivations that drive these behaviors. And this matters. Collectivist cultures generally value conformity to cultural norms, which makes prescribed norms and obligations important drivers of all behaviors, altruism included. The psychologist Toshio Yamagishi has proposed that this may explain why interpersonal trust actually tends to be lower in collectivist societies—because strong social norms make it difficult or impossible to make determinations about how trustworthy people are from their behavior. In more individualist cultures, by contrast, altruism is more likely to be motivated by individual values and choices.

There are benefits to both kinds of motivation, but one key benefit of altruism being motivated by personal values like the desire to help others is that this kind of altruism *feels* really good. An influential series of studies conducted by the psychologists Netta Weinstein and Richard Ryan found that when altruistic behavior is motivated by the altruist's personal goals and values, it results in higher levels of well-being than altruism driven by external factors. Altruism motivated by genuine compassion results not only in the glow of accomplishment and satisfaction that accompanies reaching any goal, but also in the vicarious joy that altruism promotes when it is motivated by the genuine desire to improve another's welfare. I have seen how close to the surface this joy remains for many extraordinary altruists, even long after their donations. It's the one topic that is the most likely to move them—and usually me too—to tears during our interviews. I vividly remember one altruist's description of a note he received from the mother of the boy who received his kidney. The boy's mother wrote that the kidney had started working immediately after the surgeons implanted it, and that she had never been so happy to see urine in all of her life. She added that her son was by that point off dialysis completely and had been able, for the first time in his life, to go to the beach. As she wrote, they were planning his first camping trip as well. The altruist's voice shook as he recalled receiving the letter. He told me that, to that day, "I can read it anytime and I just lose it. I just I lose it."

I have asked all of the altruists I've worked with if they would donate again, had they another kidney to spare. Every single one has said yes. Several have used the exact phrase: "In a heartbeat!" They aren't always able to articulate why, exactly, but their sentiments often echo those of one altruist who said that giving away his kidney "just satisfied something very deep within me." One particularly effusive altruist said, "If I had ten I would donate all ten. I would. It is life-changing when you do it. I cannot explain how, but your whole perception of everything changes." Another told us, "I know from my own experience there's a euphoria that accompanies the act of living donation, which is difficult to explain without sounding a little crazy. I can only liken this heightened sense of peace to the bliss many women feel after the birth of a child. It just is." Their words also remind me of the evident satisfaction felt by Lenny Skutnik, who, in interviews after his rescue, said that he "felt satisfied," because "I did what I set out to do."

Unfortunately, the joy that can accompany altruism is the source of much misunderstanding. I have been asked by more than one curious person if altruistic kidney donors are glad they donated, and if donating is something that has brought them pleasure. When I answer that, yes, they definitely are, and it does, I sometimes get a response along the lines of: "Aha! Then what they did wasn't altruistic! It was selfish, because donating made them happier!"

This is a fallacy, pure and simple. But unfortunately, it's a pervasive one. It's so pervasive that some altruists who are surprised by the pleasure their donations have brought them are left doubting their own motivations—despite all the pain and inconvenience, not to mention hundreds or thousands of dollars in financial costs, they have incurred. These doubts reflect a common but fundamental error, which is the confusion of foreseen outcomes with intended outcomes. Any goal-directed action will result in satisfaction or enjoyment when the goal is accomplished, an outcome that can be foreseen in advance. But a foreseen outcome tells us nothing about the action's *motivation*. At a psychological level, altruism is defined as acting with the ultimate goal of benefiting another's welfare, a goal that a wealth of experimental research confirms can indeed motivate both ordinary and extraordinary altruism. Whether altruists are ultimately pleased by the outcome of their actions has no bearing on this criterion.

If simply achieving pleasure was actually the goal, there are countless ways to do so that involve nowhere near the risk and discomfort of saving a stranger's life. For examples of such pleasurable activities, just ask a psychopath about how they prefer to spend their leisure time. I promise you that psychopaths, who are actually selfish and interested primarily in their own enjoyment, do not go about achieving it by helping anonymous strangers at significant risk to themselves.

Indeed, the primary reason scientists study them is because the opposite is true: they are more likely to harm strangers to benefit themselves. The fact that, for most people, alleviating others' suffering and bringing them joy can be a source of personal pleasure is, in my view, what distinguishes most of us from psychopaths—it is evidence that we *have* the capacity for genuine altruism. As the Buddhist monk and neuroscience researcher Matthieu Ricard explains, "The fact that we feel satisfaction upon completing an altruistic action presupposes that we are naturally inclined to favor the other's happiness. If we were completely indifferent to others' fates, why would we feel pleasure in taking care of them?"

That altruism can bring real pleasure is a wonderful thing. It means that engaging in altruism is *reinforcing*—the gratification it incurs makes it more likely to be repeated. What this suggests is that, if you want to be more altruistic, just start! Start small if you like. Donate blood. Register to donate bone marrow. Volunteer for a charity whose work you find meaningful. Stop to pick up something a stranger has dropped. Be spontaneous. Don't spend a lot of time mulling it over or you might end up talking yourself out of it. When you think of an achievable way to help others, one that you find meaningful, go do it. I can almost guarantee that you will be glad you did. Expending resources on helping others almost always improves well-being more than expending the same resources on yourself. And because well-being in turn increases altruism, even small acts of altruism can set into motion a virtuous cycle of giving, much as was the case for Rob Mather, whose initial fund-raising efforts on behalf of one grievously injured girl ultimately spiraled into one of the world's most powerful anti-malaria foundations.

The lives of many of the altruistic kidney donors I have worked with seem to have followed a similar trajectory: they disproportionately report having been longtime blood donors, members of the national bone marrow registry, and volunteers for charitable causes. My sense is that the gratification derived from each of these smaller acts of altruism causes some people to work their way up to extraordinary acts of altruism, "like dominoes falling," as Harold Mintz puts it. For these people, behaving altruistically has become so well learned and deeply ingrained that it has become second nature.

That the reinforcing nature of altruism can ultimately make it self-sustaining is entirely consistent with the neuroscience literature. For example, an intact maternal care system is necessary only to kick off maternal care in rodents like rats, not to keep it going. Once a mother rat has had the experience of successfully caring for offspring, even blocking all her oxytocin receptors and completely disabling the maternal care system will not affect her well-learned ability to mother. The deep-seated emotional urge to care may be a vital springboard for altruism, but once altruistic behavior has taken root, it can self-perpetuate through sheer force of habit. This may help to explain why the amygdala lesion patient S.M. is not a psychopath. She shares with psychopaths many neurocognitive deficits, including a near-total blindness to other people's fear. But whereas these deficits are lifelong in psychopaths, S.M.'s deficits were acquired—she had a partially intact amygdala until well into her teenage years. For more than a decade of her early life, she had ample opportunity to develop the habits and rewards of altruistic behavior, which seem to have been sufficiently reinforcing that she remains a kindly and generous person to this day, without any amygdala at all. This fact reinforces the critical point that the amygdala is not by itself the source of care or altruism or compassion. These are complex emergent phenomena that arise from activity in a network of interconnected brain regions. The amygdala is a vitally important node in this network, but it is only one node.

The importance of practice also helps explain why the techniques that have been empirically demonstrated to increase the capacity for altruism usually boil down to increasing opportunities for practicing it. One recent tantalizing study found that a virtual reality experience that provides people with superhero-like powers to help others may increase prosocial behavior back in the real world (or at least the laboratory). And a twenty-year-old program with demonstrated success in increasing compassionate behaviors in schoolchildren is *Roots of Empathy,* developed by Mary Gordon. The program, which provides children with the opportunity to practice caring for an infant that their classroom "adopts" for a school year, seems well designed to capitalize on children's capacities for alloparenting. During the year, children engage in caring behaviors that include recording lullabies or poems for the baby and writing down wishes they have for the baby's life. Children who complete the program are better able to recognize an infant's cries as signaling emotional distress and tend to show fewer aggressive behaviors and more everyday helping toward one another as well.

The most robustly supported way to enhance altruism appears to be via either of two related Buddhist practices called compassion and loving-kindness meditation, both of which are basically compassion boot camp. Meditators practice extending feelings of compassion, love, and generosity normally felt only for close social others to progressively more distant others, including strangers, difficult people, and ultimately all beings without distinction. The idea of experiencing genuine love and compassion for a total stranger can sound far-fetched, but just consider the fact that this kind of love is par for the course for every species of mammal on earth, humans included, in response to their babies, who, having never previously existed, are the ultimate strangers. But nearly everyone can claim the capacity to experience genuine love for such a stranger as part of our birthright. And if that capacity is there to start with, it can be built up and out. Even relatively brief training in compassion and loving-kindness meditation appears to be effective, increasing both feelings of connectedness with strangers and the tendency to behave altruistically toward them. It seems unlikely to be a coincidence that Buddhism is the dominant faith practiced by the most generous country in the world, as well as by Sunyana Graef, the first person ever to donate a kidney to an anonymous stranger, who converted to the faith as a teenager and is now a Buddhist priest. Although it's the rare major religion that doesn't advocate developing compassion toward all beings, Buddhism is unusual in having developed structured techniques for actually achieving this goal.

But meditation is only one of many routes to developing the capacity for altruism and, perhaps as important, the proclivity to use it. Several of the extraordinary altruists I have worked with are Buddhist, but most are not. Lenny Skutnik is not a Buddhist, nor is Cory Booker. Nor, to my knowledge, was the gold-toothed BMW-driving man who rescued me twenty years ago on a freeway overpass in Tacoma. All I know for certain that these people share are the capacity to recognize when another person is in need, the wherewithal to respond, and the proclivity to respond regardless of whether the person in need is someone close to them.

This suggests that before the opportunity to save someone's life confronted these people—an opportunity that, in many cases, would come to define their lives—all of them had found a way, metaphorically, to build up the earth around themselves, raising the value of even strangers' welfare to an elevated plane. The mountains on which they stand are closer to rolling hills, without large distinctions among those who populate them. In other words, all of these altruists had cultivated, wittingly or unwittingly, a sense of humility—a sense of the self as equal to and part of, rather than elevated above, all of the others around them.

That humility is the final essential ingredient for extraordinary altruism, the ingredient that binds all the others together, helps to explain what I initially found to be a puzzlingly uniform quality in extraordinary altruists: their intense resistance to all efforts to elevate them with accolades and labels like "hero." Such efforts literally make them cringe. Recall the pained smile on Cory Booker's face when a reporter asked him how he felt about being called a superhero; it reminds me exactly of how the first altruists we tested looked when some of my students fawned over them. Real heroes are truly humble people. They don't think of themselves as better or more important than other people, and being treated as though they are tends to make them extremely uncomfortable. Recall what Harold Mintz told us at the outset of our study: "Your study here is going to find out that I'm just the same as you." The fact that humility is an essential ingredient for altruism helps to explain another curious feature of extraordinary altruists: so many of them are middle-aged or older. Humility, happily, is one of those rare and wonderful qualities that tends to grow stronger and more robust from youth through adulthood and into middle age.

I started out thinking that this universal, seemingly unshakable humility, this flat-out resistance to being lionized or thought of as special in any way, was a quirky bug of altruists, but now I think it may be a feature. Humility may be, in the words of Saint Augustine, the quality that "makes men as angels." It flattens the slopes of narcissism and parochialism and ensures that truly altruistic people don't think of themselves as angels. If they did, they most likely would not be altruistic! Not only does being put on a pedestal make altruists uncomfortable, but it's not really accurate or fair to think of altruists as angels, or saints, or superheroes, or supernatural in any way. Although they are clearly more sensitive than average to others' distress, their capacity for compassion and generosity reflects the same neural mechanisms that lie latent in most of humankind. Indeed, it is in part the fact that altruists *recognize* that they are not fundamentally different from anyone else that moves them to act.

Remember this the next time you have an opportunity to help someone else, even if helping comes at some cost to yourself, even if the person in need of help is a stranger.

Remember what my roadside rescuer did.

And remember the words of Cory Booker:

"Just driving in our car, most of us see problems and challenges, and the question is: are you going to be someone who just keeps going?"

ACKNOWLEDGMENTS

This book is the product of over a decade's worth of research, and many more years' worth of reflecting on the question: why and how do humans have the capacity to care for one another? To that end, I first wish to thank the anonymous man who rescued me from a freeway overpass in Tacoma over twenty years ago, who gave me the opportunity and the incentive to work toward answering this question. I have always hoped I might one day get to thank him properly in person; perhaps I may yet.

I am also eternally grateful for the support and training of my academic mentors, who provided me with the tools and training to work toward answering this question empirically. They include my undergraduate adviser Bob Kleck, who guided me through my first research projects, and who I have always viewed as a model for what a researcher should be—unquenchably curious, open to new ideas (even those of an eager but untrained undergraduate researcher), and able to consider any question from multiple perspectives. I am also grateful to my graduate mentors, the late Nalini Ambady and Daniel Wegner, whom I admired so much and miss to this day. I learned from them many essential components of conducting psychology research, including what it means to be a supportive and encouraging mentor. I thank as well my other graduate advisers, instructors, and collaborators, including Dan Gilbert, Steven Pinker, Ken Nakayama, Nick Epley, and Bill Milberg, as well as Hillary Anger Elfenbein, Joan

Chiao, Susan Choi, Sarit Golub, Heather Gray, Meg Kozak, Jennifer Steele, and Reg Adams.

I am especially grateful to my dear friend and brilliant colleague Thalia Wheatley, whose exemplary science is equaled only by her kind heart. Among the many debts for which I can never repay her is that she introduced me to my postdoctoral adviser James Blair, who gave me the incredible opportunity to investigate the neural basis of psychopathy with him. I am grateful for his training and guidance as I navigated a learning curve that often felt like a vertical cliff, and for his modeling a genuinely interdisciplinary approach to neuroscience. How any one person can know so much about so many different topics I may never understand. The research I conducted with James also benefited from the contributions of many other wonderful NIMH scientists, including Danny Pine, Ellen Leibenluft, Gang Chen, Andreas Meyer-Lindenberg, Karina Blair, Salima Budhani, Sam Crowe, Katie Fowler, Derek Mitchell, Marina Nakic, Stu White, and Henry Yu. I am especially glad to have worked with the wise and kind Liz Finger, my indefatigable research partner during many long days, weeks, months, and years of recruiting and testing dozens of children with behavior problems. I am also very grateful to those children and their families for contributing their time and energy to our research in the hopes that what we learned from them might help other families in the future.

I am equally grateful to all the participants who have contributed to my lab's work at Georgetown. I thank all the patient and generous families who brought their children in for brain scans. My profuse thanks as well to the dozens of altruistic kidney donors who put their lives on hold to travel to Georgetown and take part in our research, and who have offered us so many invaluable insights into their experiences and decisions. I also thank our local non-kidney donor participants, without whom this research would have been impossible. Generosity comes in many forms, and their contributions of enthusiasm, time, and effort are sincerely appreciated.

I feel deeply fortunate to have had the opportunity to conduct my research at Georgetown with a group of collaborators and colleagues that includes my wonderful PhD students Elise Cardinale, Leah Lozier, Joana Vieira, Kristin Brethel-Haurwitz, Kruti Vekaria, and Katie O'Connell. Much of the science I describe is theirs every bit as much as mine. I am grateful for the many projects their brilliance and hard work have brought to fruition, and for their insight, creativity, and collaborative spirit, which make coming to work every day a genuine pleasure. Thanks also to PhD students Paul Robinson and Andrew Breeden for bringing their talents through the lab during their rotations. I have worked with many outstanding research assistants as well, including my amazing laboratory managers Sarah Stoycos, Emily Robertson, and Lydia Meena; as well as Hae Min Byeon, Jessica Chaffkin, Kelly Church, Keri Church, Michael Didow, Yean Do, Zoë Epstein, Mollie Grossman, Mike Hall, Abbey Hammell, Jenny Hammer, Alexandra Hashemi, Arianna Hughes, Sarah Khorasani, Kalli Krumpos, Kyla Machell, Diana McCue, Alissa Mrazek, Esha Nagpal, Madeleine Quinn, Nilesh Seshadri, Kelsey Smith, Madeline Smith, Maria Stoianova, Michaela Tracy, and Matt Williams.

I also thank my Georgetown colleagues John VanMeter, Rebecca Ryan, Yulia Chentsova-Dutton, Robert Veatch, and Bryce Huebner, who have contributed their expertise and wisdom to this work in myriad ways. Thank you as well to Lori Brigham at the Washington Regional Transplant Center, Reg Gohh at Brown University, Nancy Condron and Carol Williams of the Mickler's Landing Turtle Patrol, and Paula Goldberg at City Wildlife for sharing their expertise on various facets of altruism.

Thanks to the National Institute of Mental Health, the Eunice Kennedy Shriver National Institute of Child Health and Human Development, the John Templeton Foundation, and especially Marty Seligman and the Positive Neuroscience initiative for their staunch and indispensable support for my lab's research.

Of course, this book and the research it describes are not only scientific enterprises but human enterprises. And like any other human enterprise, they could never have been conceived or completed without the care and support of a great many family members and friends. From my very first research study onward, my husband Jeremy has remained my wisest counsel, strongest supporter, and best friend. He lent his fantastically expressive face to my first studies of emotion and his patience to the research they spawned. Surely any other pastime would have been more fun or interesting than reading 200 pages worth of survey responses aloud to me while I entered them into spreadsheets to complete my undergraduate thesis, but he did it with good cheer and without asking anything in return (he almost never holds this over my head anymore). And those early efforts represent a minuscule fraction of what he has contributed to my work through the years. What he has contributed to my life outside work is beyond measuring. I am grateful to be a member of a species with the capacity to love, if for no other reason than that it has given me the chance to love him.

Jeremy also gives new meaning to the word "allomother." He is a coparent without equal to our daughters, who are still too young to know how exceptional he is. But someday they will know. I am also grateful beyond words for our two daughters. Becoming their mother has made me a better person and been a source of greater joy and delight than anything I could have predicted.

One of the gifts of raising them is that it has also made me appreciate my own parents even more. If this is a story that is, in some sense, about the origins of love and care, it is a story that truly starts with my mom Margot and my dad Peter. My earliest and fondest memories are of their devoted love and care and support, and their influence shows in every cranny of this book. That I wrote it at all reflects our mutual love and enjoyment of (some might say fanaticism for) books and the written word and our thirst to better understand this miraculous world we live in. Its subject matter reflects our shared fondness for animals, faith in science to unlock the mysteries of the social and natural worlds, and belief in the possibility of human goodness. These are traits I am proud and thankful to share with them—as well as with my brother Kirt. If Kirt and I were not three years apart in age, I would be certain we were actually twins. How else to explain his uncanny ability to know what I am thinking or what I want to say before I do? His input made this book better and truer than it could otherwise have been. The same goes for the input of my awesome sister-in-law Carolyn. Of the many gifts my brother has given me, she is definitely the best. I am also grateful for the love and support of the other family members I have gained in adulthood—my own allomothers, who include my stepmother Susan and my parents-in-law Marilyn and Clark Derrick and Ron and Krista Joseph.

Lastly, I am deeply grateful to Marilia Savvides, who reached out to ask if I might like to write a book about my research—all of which she had already read and could describe to me with so much enthusiasm that she gave me the nerve to say yes. I thank her also for providing me with the opportunity to work with the thoughtful and skilled editors whose contributions to this book have been invaluable, including Hélène Barthelemy, T. J. Kelleher, and Andrew McAleer.

NOTES

Prologue

1 **the French entomologist Antoine:** Antoine Magnan, *Le Vol des insectes* (Paris: Hermann, 1934).

1 **"I applied the laws of air resistance":** Antoine Magnan, *La Locomotion Chez les Animaux*, vol. 1 (Paris: Hermann, 1934).

2 **Insects, bees included:** Douglas L. Altshuler, William B. Dickson, Jason T. Vance, Stephen P. Roberts, and Michael H. Dickinson, "Short-Amplitude High-Frequency Wing Strokes Determine the Aerodynamics of Honeybee Flight," *Proceedings of the National Academy of Sciences of the United States of America* 102, no. 50 (2005): 18213–18218.

2 **But as Charles Darwin:** Charles Darwin, *The Descent of Man* (London: John Murray Publishers, 1871).

3 **some altruism toward kin:** W. D. Hamilton, "The Genetical Evolution of Social Behaviour: I," *Journal of Theoretical Biology* 7, no. 1 (1964): 1–16; W. D. Hamilton, "The Genetical Evolution of Social Behaviour: II," *Journal of Theoretical Biology* 7, no. 1 (1964): 17–52.

3 **It explains why colony-dwelling creatures:** Paul W. Sherman, "Nepotism and the Evolution of Alarm Calls," *Science* 197, no. 4310 (1977): 1246–1253.

3 **Inclusive fitness may also explain:** Arthur J. Matas, Jodi M. Smith, Melissa A. Skeans, Kenneth E. Lamb, S. K. Gustafson, Ciara J. Samana, Darren E. Stewart, Jon J. Snyder, Ajay K. Israni, and Bertram L. Kasiske, "OPTN/SRTR 2011 Annual Data Report: Kidney," *American Journal of Transplantation* 13, suppl. 1 (2013): 11–46.

3 ***reciprocal altruism,* which relies:** Robert L. Trivers, "The Evolution of Reciprocal Altruism," *Quarterly Review of Biology* 46 (1971): 35–57.

3 **Bats are more likely to receive blood buffets:** Gerald G. Carter and Gerald S. Wilkinson, "Food Sharing in Vampire Bats: Reciprocal Help Predicts Donations More Than Relatedness or Harassment," *Proceedings of the Royal Society B: Biological Sciences* 280, no. 1753 (2013): 20122573.

4 **Some declare all altruism an illusion:** Robert B. Cialdini, Stephanie L. Brown, Brian P. Lewis, Carol Luce, et al., "Reinterpreting the Empathy-Altruism

Relationship: When One into One Equals Oneness." *Journal of Personality and Social Psychology* 73, no. 3 (1997): 481–494.

4 **Others cite supernatural forces:** Antonia J. Z. Henderson, Monica A. Landolt, Michael F. McDonald, William M. Barrable, John G. Soos, William Gourlay, Colleen J. Allison, and David N. Landsberg, "The Living Anonymous Kidney Donor: Lunatic or Saint?" *American Journal of Transplantation* 3, no. 2 (2003): 203–213.

Chapter 1: The Rescue

16 **The subjects in these studies:** Alston Chase, "Harvard and the Making of the Unabomber," *The Atlantic*, June 2000.

16 **they called "moral typecasting":** Kurt Gray and Daniel M. Wegner, "Moral Typecasting: Divergent Perceptions of Moral Agents and Moral Patients," *Journal of Personality and Social Psychology* 96, no. 3 (2009): 505–520.

17 **recognized as far back as Aristotle:** Cynthia A. Freeland, "Aristotelian Actions," *Noûs* 19, no. 3 (1985): 397–414.

19 **Social media exploded with admiration:** "Cory Booker Hashtag Explodes on Twitter After Mayor's Dramatic Fire Rescue," *NJ.com*, April 13, 2012, http://www.nj.com/news/index.ssf/2012/04/cory_booker_hashtag_explodes_o.html; Stephanie Haberman, "Super-Mayor Cory Booker Gets Memed," *CNN.com*, April 13, 2012, http://www.cnn.com/2012/04/13/tech/web/mayor-cory-booker-memed/.

19 **Booker was blunt:** "Mayor Cory Booker Answers Questions About Fire Rescue," uploaded April 13, 2012, https://www.youtube.com/watch?v=Wm1SKTyVZf8; "Newark Mayor Cory Booker: Race into Home Fire Was 'a Come to Jesus Moment,'" *CBS News*, April 13, 2012, http://www.cbsnews.com/news/newark-mayor-cory-booker-race-into-home-fire-was-a-come-to-jesus-moment/; Alyssa Newcomb, "Newark Mayor Cory Booker Rescues Neighbor from Fire," *ABC News*, April 13, 2012, http://abcnews.go.com/blogs/headlines/2012/04/newark-mayor-cory-booker-rescues-neighbor-from-fire/; James Barron, "After Rescuing Woman from Fire, a Mayor Recalls His Fear and Focus," *New York Times*, April 13, 2012.

Chapter 2: Heroes and Antiheroes

23 **he'd landed on his head:** "Ex-Stanford Wrestler Surmon Dies in Fall," *Washington Post*, January 3, 2000.

24 **More than one serial killer:** "Suspected or Convicted Serial Killers in Washington," *Seattle Post-Intelligencer*, February 19, 2003; "Adhahn Pleads Guilty to Murder of Tacoma Girl, 12," *Seattle Times*, April 8, 2008.

24 **John Keats was correct in observing:** John Keats, letter to George and Georgiana Keats, February 14–May 3, 1819, http://www.john-keats.com/briefe/140219.htm.

25 **the infamous case of Kitty Genovese:** Rachel Manning, Mark Levine, and Alan Collins, "The Kitty Genovese Murder and the Social Psychology of Helping: The Parable of the 38 Witnesses," *American Psychologist* 62, no. 6 (2007): 555–562.

25 **follow-up psychology studies:** John M. Darley and Bibb Latané, "Bystander Intervention in Emergencies: Diffusion of Responsibility," *Journal of Personality and Social Psychology* 8, no. 4 (1968): 377–383.

25 **Zimbardo's infamous Stanford Prison Experiment:** Craig Haney, Curtis Banks, and Philip Zimbardo, "Interpersonal Dynamics in a Simulated Prison," *International Journal of Criminology and Penology* 1 (1973): 69–97.

26 **Stanley Milgram's research was so controversial:** Daniel Raver, "Stanley Milgram," Psyography, http://faculty.frostburg.edu/mbradley/psyography/stanleymilgram.html.

26 **most influential psychologists of the last century:** Ibid.

26 **"six degrees of separation" is a real thing:** Jeffrey Travers and Stanley Milgram, "An Experimental Study of the Small World Problem," *Sociometry* (1969): 425–443.

26 **the most notorious use ever of electric shocks:** For original documentation of the studies, see Stanley Milgram, "Behavioral Study of Obedience," *Journal of Abnormal Psychology* 67 (1963): 371–378; Stanley Milgram, "Obedience" (film) (New York: New York University Film Library, 1965).

28 **he anxiously monitored the local obituaries:** Sandi Kahn Shelton, "Clinton Man Hears Dad's Taped Screams in 1960s Shock Study at Yale" (video), *New Haven Register,* October 20, 2012.

28 **"There is a need to draw a line":** Isabel Kershner, "Pardon Plea by Adolf Eichmann, Nazi War Criminal, Is Made Public," *New York Times,* January 27, 2016.

29 **more recent evidence has suggested:** S. Alexander Haslam and Stephen D. Reicher, "Contesting the 'Nature' of Conformity: What Milgram and Zimbardo's Studies Really Show," *PLoS Biology* 10, no. 11 (2012): e1001426.

30 **"I would say, on the basis":** Gina Perry, *Behind the Shock Machine: The Untold Story of the Notorious Milgram Psychology Experiments* (New York: New Press, 2013).

31 **Versions of the study have been run:** Thomas Blass, "The Milgram Paradigm After 35 Years: Some Things We Now Know About Obedience to Authority," *Journal of Applied Social Psychology* 29, no. 5 (1999): 955–978; Thomas Blass, "A Cross-Cultural Comparison of Studies of Obedience Using the Milgram Paradigm: A Review," *Social and Personality Psychology Compass* 6, no. 2 (2012): 196–205; Jerry M. Burger, "Replicating Milgram: Would People Still Obey Today?" *American Psychologist* 64, no. 1 (2009): 1–11.

31 **the volunteers' responses weren't uniform:** Haslam and Reicher, "Contesting the 'Nature' of Conformity."

34 **Batson's study used electrical shocks:** C. Daniel Batson, Bruce D. Duncan, Paula Ackerman, Terese Buckley, and Kimberly Birch, "Is Empathic Emotion a Source of Altruistic Motivation?" *Journal of Personality and Social Psychology* 40, no. 2 (1981): 290–302.

39 **a discipline that has historically focused:** Richard Nisbett and Lee Ross, *The Person and the Situation: Perspectives of Social Psychology* (New York: McGraw-Hill, 1991). For a historical discussion of this perspective, see Harry T. Reis, "Reinvigorating the Concept of Situation in Social Psychology," *Personality and Social Psychology Review* 12, no. 4 (2008): 311–329.

40 **So another possibility—one of many:** Catherine Tuvblad and Laura A. Baker, "Human Aggression Across the Lifespan: Genetic Propensities and Environmental Moderators," *Advances in Genetics* 75 (2011): 171–214.

40 **Many of these studies aren't designed:** For a review, see Robert Plomin, "Genetics and Developmental Psychology," *Merrill-Palmer Quarterly* 50, no. 3 (2004): 341–352. For one example of a study that examines the genetic influences of an effect that is often described as purely environmental—the influence of parents' language use on children's language development—see Laura S. DeThorne, Stephen A. Petrill, Sara A. Hart, Ron W. Channell, Rebecca J. Campbell, Kirby Deater-Deckard, Lee Anne Thompson, and David J. Vandenbergh, "Genetic

Effects on Children's Conversational Language Use," *Journal of Speech, Language, and Hearing Research* 51, no. 2 (2008): 423–435.

41 **A study like this provides compelling evidence:** Thomas J. Bouchard Jr., David T. Lykken, Matthew McGue, Nancy L. Segal, and Auke Tellegen, "Sources of Human Psychological Differences: The Minnesota Study of Twins Reared Apart," *Science* 250, no. 4978 (1990): 223–228; Robert Plomin, Michael J. Owen, and Peter McGuffin, "The Genetic Basis of Complex Human Behaviors," *Science* 264, no. 5166 (1994): 1733–1739. Some aspects of genetic contributions to human traits remain topics of debate; for a discussion, see Eric Turkheimer, "Three Laws of Behavior Genetics and What They Mean," *Current Directions in Psychological Science* 9, no. 5 (2000): 160–164.

42 **This is how researchers could determine:** Turi E. King, Gloria Gonzalez Fortes, Patricia Balaresque, Mark G. Thomas, David Balding, Pierpaolo Maisano Delser, Rita Neumann, Walther Parson, Michael Knapp, and Susan Walsh, "Identification of the Remains of King Richard III," *Nature Communications* 5 (2014): 5631.

42 **A caveat is that the heritability:** For a review of heritability and misconceptions about it, see Peter M. Visscher, William G. Hill, and Naomi R. Wray, "Heritability in the Genomics Era: Concepts and Misconceptions," *Nature Reviews Genetics* 9, no. 4 (2008): 255–266. For important findings relevant to gene-environment interactions, see Eric Turkheimer, Andreana Haley, Mary Waldron, Brian D'Onofrio, and Irving I. Gottesman, "Socioeconomic Status Modifies Heritability of IQ in Young Children," *Psychological Science* 14, no. 6 (2003): 623–628; Elliot M. Tucker-Drob and Timothy C. Bates, "Large Cross-National Differences in Gene × Socioeconomic Status Interaction on Intelligence," *Psychological Science* 27, no. 2 (2015): 138–149. For an examination of how genetic and environmental processes shape human height variation, see Gert Stulp and Louise Barrett, "Evolutionary Perspectives on Human Height Variation," *Biological Reviews of the Cambridge Philosophical Society* 91, no. 1 (2016): 206–234.

42 **For body weight, heritability hovers:** Hermine H. M. Maes, Michael C. Neale, and Lindon J. Eaves, "Genetic and Environmental Factors in Relative Body Weight and Human Adiposity," *Behavior Genetics* 27, no. 4 (1997): 325–351.

43 **the famed behavioral geneticist Eric Turkheimer:** Turkheimer, "Three Laws of Behavior Genetics."

43 **A massive study reported in the journal:** Tinca J. C. Polderman, Beben Benyamin, Christiaan A. de Leeuw, Patrick F. Sullivan, Arjen van Bochoven, Peter M. Visscher, and Danielle Posthuma, "Meta-Analysis of the Heritability of Human Traits Based on Fifty Years of Twin Studies," *Nature Genetics* 47, no. 7 (2015): 702–709.

43 **In Milgram's era, most psychologists:** Plomin, "Genetics and Developmental Psychology." For an in-depth review of this topic, see the outstanding book by Steven Pinker, *The Blank Slate: The Modern Denial of Human Nature* (New York: Penguin, 2003).

44 **The behaviorists' views were very influential:** Steven J. Haggbloom, Renee Warnick, Jason E. Warnick, Vinessa K. Jones, Gary L. Yarbrough, Tenea M. Russell, Chris M. Borecky, Reagan McGahhey, John L. Powell III, and Jamie Beavers, "The 100 Most Eminent Psychologists of the 20th Century," *Review of General Psychology* 6, no. 2 (2002): 139.

44 **"What is love except":** Burrhus Frederic Skinner, *Walden Two* (Indianapolis: Hackett Publishing, 1974) 282.

45 if we could perfectly control: Burrhus Frederic Skinner, *Science and Human Behavior* (New York: Simon & Schuster, 1953).

45 The heritability of aggression is: Tuvblad and Baker, "Human Aggression Across the Lifespan"; Marina A. Bornovalova, Brian M. Hicks, William G. Iacono, and Matt McGue, "Familial Transmission and Heritability of Childhood Disruptive Disorders," *American Journal of Psychiatry* 167, no. 9 (2010): 1066–1074; S. Alexandra Burt, "Are There Meaningful Etiological Differences Within Antisocial Behavior? Results of a Meta-Analysis," *Clinical Psychology Review* 29 (2009): 163–178; Laura A. Baker, Adrian Raine, Jianghong Liu, and Kristen C. Jacobson, "Differential Genetic and Environmental Influences on Reactive and Proactive Aggression in Children," *Journal of Abnormal Child Psychology* 36, no. 8 (2008): 1265–1278; Dehryl A. Mason and Paul J. Frick, "The Heritability of Antisocial Behavior: A Meta-Analysis of Twin and Adoption Studies," *Journal of Psychopathology and Behavioral Assessment* 16, no. 4 (1994): 301–323.

45 but among violent criminals: Robert D. Hare, *Without Conscience: The Disturbing World of the Psychopaths Among Us* (New York: Guilford, 1993).

46 Psychopathy is also highly influenced: Catherine Tuvblad, Serena Bezdjian, Adrian Raine, and Laura A. Baker, "The Heritability of Psychopathic Personality in 14- to 15-Year-Old Twins: A Multirater, Multimeasure Approach," *Psychological Assessment* 26 (2014): 704–716; Essi Viding, Robert James R. Blair, Terrie E. Moffitt, and Robert Plomin, "Evidence for Substantial Genetic Risk for Psychopathy in 7-Year-Olds," *Journal of Child Psychology and Psychiatry* 46, no. 6 (2005): 592–597.

46 Not so for Gary, who grew up: Ann Rule, *Green River, Running Red: The Real Story of the Green River Killer—America's Deadliest Serial Murderer* (New York: Simon & Schuster, 2004); Terry McCarthy, "River of Death," *Time* 159, no. 22 (2002): 56–61; Mary Ellen O'Toole and Alisa Bowman, *Dangerous Instincts: How Gut Feelings Betray Us* (New York: Hudson Street Press, 2011).

48 Tony Savage emphasized this: "Inside the Mind of Serial Killer Gary Ridgway," *Larry King Live*, CNN, February 18, 2004, http://www.cnn.com/TRANSCRIPTS/0402/18/lkl.00.html.

48 Most people who are psychotic: Sherry Glied and Richard G. Frank, "Mental Illness and Violence: Lessons from the Evidence," *American Journal of Public Health* 104, no. 2 (2014): e5–e6.

50 Children who are abused or neglected: David D. Vachon, Robert F. Krueger, Fred A. Rogosch, and Dante Cicchetti, "Assessment of the Harmful Psychiatric and Behavioral Effects of Different Forms of Child Maltreatment," *JAMA Psychiatry* (2015): 1135–1142; Gayla Margolin and Elana B. Gordis, "The Effects of Family and Community Violence on Children," *Annual Review of Psychology* 51 (2000): 445–479.

50 what really sets them apart is *proactive* aggression: Robert James R. Blair, "Neurocognitive Models of Aggression, the Antisocial Personality Disorders, and Psychopathy," *Journal of Neurology, Neurosurgery, and Psychiatry* 71, no. 6 (2001): 727–731.

50 It's not like people haven't looked: K. A. Dodge, John E. Lochman, Jennifer D. Harnish, John E. Bates, and G. S. Pettit, "Reactive and Proactive Aggression in School Children and Psychiatrically Impaired Chronically Assaultive Youth," *Journal of Abnormal Psychology* 106, no. 1 (1997): 37; Julian D. Ford, Lisa A. Fraleigh, and Daniel F. Connor, "Child Abuse and Aggression Among Seriously

Emotionally Disturbed Children," *Journal of Clinical Child and Adolescent Psychology* 39, no. 1 (2010): 25–34.

50 **one large study conducted by Adrian Raine:** Catherine Tuvblad, Adrian Raine, Mo Zheng, and Laura A. Baker, "Genetic and Environmental Stability Differs in Reactive and Proactive Aggression," *Aggressive Behavior* 35, no. 6 (2009): 437–452.

53 **one of my dissertation studies:** Abigail A. Marsh, Megan N. Kozak, and Nalini Ambady, "Accurate Identification of Fear Facial Expressions Predicts Prosocial Behavior," *Emotion* 7, no. 1 (2007): 239–251; Jay S. Coke, C. Daniel Batson, and Katherine McDavis, "Empathic Mediation of Helping: A Two-Stage Model," *Journal of Personality and Social Psychology* 36 (1978): 752–766.

55 **one of the "most unintuitive" psychology findings:** Simon A. Moss and Samuel Wilson, "Integrating the Most Unintuitive Empirical Observations of 2007 in the Domain of Personality and Social Psychology into a Unified Framework," *New Ideas in Psychology* 28, no. 1 (2010): 1–27.

55 **Subsequent research has also linked:** Abigail A. Marsh and R. J. Blair, "Deficits in Facial Affect Recognition Among Antisocial Populations: A Meta-Analysis," *Neuroscience and Biobehavioral Reviews* 32 (2008): 454–465; Purva Rajhans, Nicole Altvater-Mackensen, Amrisha Vaish, and Tobias Grossmann, "Children's Altruistic Behavior in Context: The Role of Emotional Responsiveness and Culture," *Scientific Reports* 6 (2016): 24089; Abigail A. Marsh, Sarah A. Stoycos, Kristin M. Brethel-Haurwitz, Paul Robinson, and Elise M. Cardinale, "Neural and Cognitive Characteristics of Extraordinary Altruists," *Proceedings of the National Academy of Sciences* 111, no. 42 (2014): 15036–15041; Stuart F. White, Margaret J. Briggs-Gowan, Joel L. Voss, Amelie Petitclerc, Kimberly McCarthy, R. J. R. Blair, and Laurie S. Wakschlag, "Can the Fear Recognition Deficits Associated with Callous-Unemotional Traits Be Identified in Early Childhood?" *Journal of Clinical and Experimental Neuropsychology* 38, no. 6 (2016): 672–684.

Chapter 3: The Psychopathic Brain

58 **James proposed that the mechanism:** R. J. Blair, "A Cognitive Developmental Approach to Morality: Investigating the Psychopath," *Cognition* 57, no. 1 (1995): 1–29; R. J. Blair, "Applying a Cognitive Neuroscience Perspective to the Disorder of Psychopathy," *Development and Psychopathology* 17, no. 3 (2005): 865–891.

59 **the work of animal behavior experts:** Konrad Lorenz, *On Aggression* (London: Methuen, 1966); Irenäus Eibl-Eibesfeldt, *Love and Hate: The Natural History of Behavior Patterns* (Chicago: Aldine, 1996); Rudolf Schenkel, "Submission: Its Features in the Wolf and Dog," *American Zoologist* 7 (1967): 319–329.

61 **Young children are almost always:** Sylvana M. Côté, Tracy Vaillancourt, John C. LeBlanc, Daniel S. Nagin, and Richard E. Tremblay, "The Development of Physical Aggression from Toddlerhood to Pre-adolescence: A Nation Wide Longitudinal Study of Canadian Children," *Journal of Abnormal Child Psychology* 34, no. 1 (2006): 71–85; Richard E. Tremblay, "The Development of Physical Aggression," Encyclopedia on Early Childhood Development, January 2012, http://www.child-encyclopedia.com/aggression/according-experts/development-physical-aggression.

61 **In one study conducted in the 1970s:** Linda A. Camras, "Facial Expressions Used by Children in a Conflict Situation," *Child Development* (1977): 1431–1435.

62 **even during negotiations between adults:** Marwan Sinaceur, Shirli Kopelman, Dimitri Vasiljevic, and Christopher Haag, "Weep and Get More: When and Why Sadness Expression Is Effective in Negotiations," *Journal of Applied Psychology* 100, no. 6 (2015): 1847–1871.

62 **Here are the full criteria for a conduct disorder diagnosis:** *Diagnostic and Statistical Manual of Mental Disorders* (Washington, DC: American Psychiatric Association, 2013), 469–471.

64 **these children may not respond appropriately:** Robert James R. Blair, "Responsiveness to Distress Cues in the Child with Psychopathic Tendencies," *Personality and Individual Differences* 27, no. 1 (1999): 135–145; Amy Dawel, Richard O'Kearney, Elinor McKone, and Romina Palermo, "Not Just Fear and Sadness: Meta-Analytic Evidence of Pervasive Emotion Recognition Deficits for Facial and Vocal Expressions in Psychopathy," *Neuroscience and Biobehavioral Reviews* 36 (2012): 2288–2304; Abigail A. Marsh and R. J. R. Blair, "Deficits in Facial Affect Recognition Among Antisocial Populations: A Meta-Analysis," *Neuroscience and Biobehavioral Reviews* 32 (2008): 454–465; Stuart F. White, Margaret J. Briggs-Gowan, Joel L. Voss, Amelie Petitclerc, Kimberly McCarthy, R. James R. Blair, and Lauren S. Wakschlag, "Can the Fear Recognition Deficits Associated with Callous-Unemotional Traits Be Identified in Early Childhood," *Journal of Clinical and Experimental Neuropsychology* 38, no. 6 (2016): 672–684; Patrick D. Sylvers, Patricia A. Brennan, and Scott O. Lilienfeld, "Psychopathic Traits and Preattentive Threat Processing in Children: A Novel Test of the Fearlessness Hypothesis," *Psychological Science* 22 (2011): 1280–1287.

66 **When electricity is sent surging:** José Manuel Rodríguez Delgado, *Physical Control of the Mind: Toward a Psychocivilized Society* (Washington, DC: World Bank Publications, 1969); Marvin Wasman and John P. Flynn, "Directed Attack Elicited from Hypothalamus," *Archives of Neurology* 6 (1962): 220–227; Thomas R. Gregg and Allan Siegel, "Brain Structures and Neurotransmitters Regulating Aggression in Cats: Implications for Human Aggression," *Progress in Neuro-Psychopharmacology and Biological Psychiatry* 25, no. 1 (2001): 91–140.

66 **Optogenetic triggering of neurons:** Dayu Lin, Maureen P. Boyle, Piotr Dollar, Hyosang Lee, E. S. Lein, Pietro Perona, and David J. Anderson, "Functional Identification of an Aggression Locus in the Mouse Hypothalamus," *Nature* 470, no. 7333 (2011): 221–226..

66 **electrically induced rage will not be directed:** Alexander M. Perachio, "The Influence of Target Sex and Dominance on Evoked Attack in Rhesus Monkeys," *American Journal of Physical Anthropology* 38, no. 2 (1973): 543–547.

69 **The modern clinical definition of psychopathy:** Hervey Cleckley, *The Mask of Sanity: An Attempt to Clarify Some Issues About the So-called Psychopathic Personality* (Brattleboro, VT: Echo Point Books & Media, 2015).

69 **More often than not, the typical psychopath:** Ibid., 339.

70 **Perhaps our resistance to the idea:** Although the perception of children as moral patients may be less true for black children, who tend to be perceived as older and physically larger and more likely to be guilty of a crime than white children of the same age; see, for example, Phillip Atiba Goff, Matthew Christian Jackson, Brooke Allison, Lewis Di Leone, Carmen Marie Culotta, and Natalie Ann DiTomasso, "The Essence of Innocence: Consequences of Dehumanizing Black Children," *Journal of Personality and Social Psychology* 106, no. 4 (2014): 526–545, DOI: 10.1037/a0035663; American Psychological Association, "Black

Boys Viewed as Older, Less Innocent Than Whites, Research Finds," March 6, 2014, http://www.apa.org/news/press/releases/2014/03/black-boys-older.aspx.

71 **The title of a widely circulated:** Jennifer Kahn, "Can You Call a 9-Year-Old a Psychopath?" *New York Times Magazine,* May 11, 2012.

71 **There is a nearly identical 40-point scale:** Robert D. Hare, *The Hare Psychopathy Checklist—Revised* (Toronto: Multi-Health Systems, 1991); Adelle E. Forth, David S. Kosson, and Robert D. Hare, *The Psychopathy Checklist: Youth Version* (Toronto: Multi-Health Systems, 2003).

76 **in children with psychopathic traits the opposite:** Ashley S. Hampton, Deborah A. G. Drabick, and Laurence Steinberg, "Does IQ Moderate the Relation Between Psychopathy and Juvenile Offending?" *Law and Human Behavior* 38, no. 1 (2014): 23–33; Terrie E. Moffitt, "Adolescence-Limited and Life-Course-Persistent Antisocial Behavior: A Developmental Taxonomy," *Psychological Review* 100, no. 4 (1993): 674–701.

79 **nearly all of these parents had other children:** For Ridgeway's brother's response to reading about Ridgway's crimes, see Michael Ko, "Ridgway's Relatives 'Mortified by Grief,'" *Seattle Times,* November 9, 2003.

80 **different styles of parenting may buffer:** Hugh Lytton, "Child and Parent Effects in Boys' Conduct Disorder: A Reinterpretation," *Developmental Psychology* 26, no. 5 (1990): 683–697; Grazyna Kochanska, "Multiple Pathways to Conscience for Children with Different Temperaments: From Toddlerhood to Age 5," *Developmental Psychology* 33, no. 2 (1997): 228–240; Rebecca Waller, Frances Gardner, Essi Viding, Daniel S. Shaw, Thomas J. Dishion, Melvin N. Wilson, and Luke W. Hyde, "Bidirectional Associations Between Parental Warmth, Callous Unemotional Behavior, and Behavior Problems in High-Risk Preschoolers," *Journal of Abnormal Child Psychology* 42, no. 8 (2014): 1275–1285; Rebecca Waller, Daniel S. Shaw, Erika E. Forbes, and Luke W. Hyde, "Understanding Early Contextual and Parental Risk Factors for the Development of Limited Prosocial Emotions," *Journal of Abnormal Child Psychology* 43, no. 6 (2015): 1025–1039.

80 **very high levels of parental warmth:** Luke W. Hyde, Rebecca Waller, Christopher J. Trentacosta, Daniel S. Shaw, Jenae M. Neiderhiser, Jody M. Ganiban, David Reiss, and Leslie D. Leve, "Heritable and Nonheritable Pathways to Early Callous-Unemotional Behaviors," *American Journal of Psychiatry* 173, no. 9 (2016): 903–910; Bruce Rosen, "fMRI at 20: Has It Changed the World?" ISMRM Lauterbur Lecture, 2011, uploaded April 9, 2013, https://www.youtube.com/watch?v=edO43AT5GhE.

81 **the NIMH acquired a 7-Tesla magnet:** Jun Shen, "Section on Magnetic Resonance Spectroscopy," National Institute of Mental Health, https://www.nimh.nih.gov/labs-at-nimh/research-areas/clinics-and-labs/mib/smrs/index.shtml; Björn Friebe, Astrid Wollrab, Markus Thormann, Katharina Fischbach, Jens Ricke, Marcus Grueschow, Siegfried Kropf, Frank Fischbach, and Oliver Speck, "Sensory Perceptions of Individuals Exposed to the Static Field of a 7T MRI: A Controlled Blinded Study," *Journal of Magnetic Resonance Imaging* 41, no. 6 (2015): 1675–1681.

82 **a group of researchers, led by Daniel Tranel:** Daniel Tranel and Bradley T. Hyman, "Neuropsychological Correlates of Bilateral Amygdala Damage," *Archives of Neurology* 47, no. 3 (1990): 349–355.

83 **photographs of people who looked frightened:** Ralph Adolphs, Daniel Tranel, Hanna Damasio, and Antonio Damasio, "Impaired Recognition of Emotion in

Facial Expressions Following Bilateral Damage to the Human Amygdala," *Nature* 372, no. 6507 (1994): 669–672.

83 **revealed her knack for portraiture:** Ralph Adolphs, Daniel Tranel, Hanna Damasio, and Antonio R. Damasio, "Fear and the Human Amygdala," *Journal of Neuroscience* 15, no. 9 (1995): 5879–5891.

83 **in a teenage Urbach-Wiethe patient:** Morteza Pishnamazi, Abbas Tafakhori, Sogol Loloee, Amirhossein Modabbernia, Vajiheh Aghamollaii, Bahador Bahrami, and Joel S. Winston, "Attentional Bias Towards and Away from Fearful Faces Is Modulated by Developmental Amygdala Damage," *Cortex* 81 (2016): 24–34.

83 **including vocal utterances, body postures:** For a review, see Abigail A. Marsh, "Understanding Amygdala Responsiveness to Fearful Expressions Through the Lens of Psychopathy and Altruism," *Journal of Neuroscience Research* 94, no. 6 (2016): 513–525.

84 **Hundreds of studies have now been conducted:** Paolo Fusar-Poli, Anna Placentino, Francesco Carletti, Paola Landi, Paul Allen, Simon Surguladze, Francesco Benedetti, Marta Abbamonte, Roberto Gasparotti, Francesco Barale, Jorge Perez, Philip McGuire, and Pierliugi Politi, "Functional Atlas of Emotional Faces Processing: A Voxel-Based Meta-Analysis of 105 Functional Magnetic Resonance Imaging Studies," *Journal of Psychiatry and Neuroscience* 34, no. 6 (2009): 418–432.

85 **A firefighter once got sucked into an MRI:** J. K. Bucsio, "MRI Facility Safety: Understanding the Risks of Powerful Attraction," *Radiology Today* 6, no. 22 (2005): 22.

86 **The simple act of labeling an emotion:** Matthew D. Lieberman, Naomi I. Eisenberger, Molly J. Crockett, Sabrina M. Tom, Jennifer H. Pfeifer, and Baldwin M. Way, "Putting Feelings into Words: Affect Labeling Disrupts Amygdala Activity in Response to Affective Stimuli," *Psychological Science* 18, no. 5 (2007): 421–428.

88 **the region of the brain that is critical:** Abigail A. Marsh, Elizabeth C. Finger, Derek G. V. Mitchell, Marguerite E. Reid, Courtney Sims, David S. Kosson, Kenneth E. Towbin, Ellen Leibenluft, Daniel S. Pine, and R. James R. Blair, "Reduced Amygdala Response to Fearful Expressions in Children and Adolescents with Callous-Unemotional Traits and Disruptive Behavior Disorders," *American Journal of Psychiatry* 165, no. 6 (2008): 712–720; Alice P. Jones, Kristin R. Laurens, Catherine M. Herba, Gareth J. Barker, and Essi Viding, "Amygdala Hypoactivity to Fearful Faces in Boys with Conduct Problems and Callous-Unemotional Traits," *American Journal of Psychiatry* 166 (2009): 95–102; Essi Viding, Catherine L. Sebastian, Mark R. Dadds, Patricia L. Lockwood, Charlotte A. M. Cecil, Stephane A. De Brito, and Eamon J. McCrory, "Amygdala Response to Preattentive Masked Fear in Children with Conduct Problems: The Role of Callous-Unemotional Traits," *American Journal of Psychiatry* 169, no. 10 (2012): 1109–1116; Stuart F. White, Abigail A. Marsh, Katherine A. Fowler, Julia C. Schechter, Christopher Adalio, Kayla Pope, Stephen Sinclair, Daniel S. Pine, and R. James R. Blair, "Reduced Amygdala Response in Youths with Disruptive Behavior Disorders and Psychopathic Traits: Decreased Emotional Response Versus Increased Top-Down Attention to Nonemotional Features," *American Journal of Psychiatry* 169, no. 7 (2012): 750–758; Leah M. Lozier, Elise M. Cardinale, John W. VanMeter, and Abigail A. Marsh, "Mediation of the Relationship Between Callous-Unemotional Traits and Proactive Aggression by Amygdala Response to Fear Among Children with Conduct Problems," *JAMA Psychiatry* 71, no. 6 (2014): 627–636.

90 **Ours was not the only study:** Abigail A. Marsh, Elizabeth E. Finger, Julia C. Schechter, Ilana T. N. Jurkowitz, Marguerite Reid Schneider, and Robert James R. Blair, "Adolescents with Psychopathic Traits Report Reductions in Physiological Responses to Fear," *Journal of Child Psychology and Psychiatry* 52, no. 8 (2011): 834–841; Alice Jones Bartoli, Francesca G. Happe, Francesca Gilbert, Stephanie Burnett Heyes, and Essi Viding, "Feeling, Caring, Knowing: Different Types of Empathy Deficit in Boys with Psychopathic Tendencies and Autism Spectrum Disorder," *Journal of Child Psychology and Psychiatry* 51, no. 11 (2010): 1188–1197; Rebecca Waller, Christopher J. Trentacosta, Daniel S. Shaw, Jenae M. Neiderhiser, Jody M. Ganiban, David Reiss, Leslie D. Leve, and Luke W. Hyde, "Heritable Temperament Pathways to Early Callous-Unemotional Behaviour," *British Journal of Psychiatry* 209, no. 6 (2016): 475–482; Ida Klingzell, Kostas A. Fanti, Olivier F. Colins, Louise Frogner, Anna-Karin Andershed, and Henrik Andershed, "Early Childhood Trajectories of Conduct Problems and Callous-Unemotional Traits: The Role of Fearlessness and Psychopathic Personality Dimensions," *Child Psychiatry and Human Development* 47, no. 2 (2016): 236–247; Kostas A. Fanti, Georgia Panayiotou, C. Lazarou, R. Michael, and Giorgos Georgiou, "The Better of Two Evils? Evidence That Children Exhibiting Continuous Conduct Problems High or Low on Callous-Unemotional Traits Score on Opposite Directions on Physiological and Behavioral Measures of Fear," *Development and Psychopathology* 28, no. 1 (2016): 185–198.

90 **Similar fearlessness has been observed:** Justin S. Feinstein, Ralph Adolphs, Antonio Damasio, and Daniel Tranel, "The Human Amygdala and the Induction and Experience of Fear," *Current Biology* 21 (2011): 34–38; Justin S. Feinstein, "Lesion Studies of Human Emotion and Feeling," *Current Opinion in Neurobiology* 23, no. 3 (2013): 304–309: Michael Davis, "The Role of the Amygdala in Fear and Anxiety," *Annual Review of Neuroscience* 15 (1992): 353–75.

91 **As one psychopathic sex offender interviewed:** Robert D. Hare, *Without Conscience: The Disturbing World of the Psychopaths Among Us* (New York: Guilford, 1993), 44.

91 **these aberrant judgments correspond:** Abigail A. Marsh and Elise M. Cardinale, "Psychopathy and Fear: Specific Impairments in Judging Behaviors That Frighten Others," *Emotion* 12, no. 5 (2012): 892–898; Abigail A. Marsh and Elise M. Cardinale, "When Psychopathy Impairs Moral Judgments: Neural Responses During Judgments About Causing Fear," *Social Cognitive and Affective Neuroscience* 9 (2014): 3–11; Elise M. Cardinale and Abigail A. Marsh, "Impact of Psychopathy on Moral Judgments About Causing Fear and Physical Harm," *PLoS One* 10, no. 5 (2015): e0125708.

Chapter 4: The Other Side of the Curve

94 **my student Joana Vieira and others:** Yaling Yang, Adrian Raine, Katherine L. Narr, Patrick M. Colletti, and Arthur W. Toga, "Localization of Deformations Within the Amygdala in Individuals with Psychopathy," *Archives of General Psychiatry* 66, no. 9 (2009): 986–994; Dustin A. Pardini, Adrian Raine, Kurt Erickson, and Rolf Loeber, "Lower Amygdala Volume in Men Is Associated with Childhood Aggression, Early Psychopathic Traits, and Future Violence," *Biological Psychiatry* 75, no. 1 (2014): 73–80; Joana B. Vieira, Fernando Ferreira-Santos, Pedro R. Almeida, F. Barbosa, João Marques-Teixeira, and Abigail A. Marsh,

"Psychopathic Traits Are Associated with Cortical and Subcortical Volume Alterations in Healthy Individuals," *Social Cognitive and Affective Neuroscience* 10, no. 12 (2015): 1693–1704; Moran D. Cohn, Essi Viding, Eamon McCrory, Louise Pape, Wim van den Brink, Theo A. H. Doreleijers, Dick J. Veltman, and Arne Popma, "Regional Grey Matter Volume and Concentration in At-Risk Adolescents: Untangling Associations With Callous-Unemotional Traits and Conduct Disorder Symptoms," *Psychiatry Research* 254 (2016): 180–87.

94 **a 2014 study of amygdala activity:** Leah M. Lozier, Elise M. Cardinale, John W. VanMeter, and Abigail A. Marsh, "Mediation of the Relationship Between Callous-Unemotional Traits and Proactive Aggression by Amygdala Response to Fear Among Children with Conduct Problems," *JAMA Psychiatry* 71, no. 6 (2014): 627–636.

99 **perhaps 30 percent of the population registers:** Jeremy Coid, Min Yang, Simone Ullrich, Amanda Roberts, and Robert D. Hare, "Prevalence and Correlates of Psychopathic Traits in the Household Population of Great Britain," *International Journal of Law and Psychiatry* 32, no. 2 (2009): 65–73.

100 **the remaining 61 percent were generous:** Ziv G. Epstein, Alexander Peysakhovich, and David G. Rand, "The Good, the Bad, and the Unflinchingly Selfish: Cooperative Decision-Making Can Be Predicted with High Accuracy Using Only Three Behavioral Types," paper presented to the Seventeenth ACM Conference on Economics and Computation, June 3, 2016, https://ssrn.com/abstract=2737983.

100 **"All the friendly feelings are derived":** Aristotle, *Ethics*, part 5, http://www.fullbooks.com/Ethics5.html.

100 **belief that human nature is fundamentally selfish:** Dale T. Miller, "The Norm of Self-Interest," *American Psychologist* 54, no. 12 (1999): 1053–1060.

101 **A nearly identical percentage of people:** Robert Wuthnow, *Acts of Compassion: Caring for Others and Helping Ourselves* (Princeton, NJ: Princeton University Press, 1991). See results of *New York Times/CBS News* poll, conducted July 17–19, 1999, at: https://partners.nytimes.com/library/national/101799mag-poll-results.html. To explore the GSS data, visit GSS Data Explorer, "Can People Be Trusted," https://gssdataexplorer.norc.org/variables/441/vshow.

102 **Most traits, from height to cholesterol levels:** For population distributions, see Centers for Disease Control and Prevention (CDC), "Anthropometric Reference Data for Children and Adults: United States, 2007–2010," *Vital and Health Statistics* 11, no. 252 (October 2012), https://www.cdc.gov/nchs/data/series/sr_11/sr11_252.pdf.

102 **what is called a *half normal curve*:** Coid et al., "Prevalence and Correlates of Psychopathic Traits."

104 **People with this condition represent:** Jeremy B. Wilmer, Laura Germine, Christopher F. Chabris, Garga Chatterjee, Mark Williams, Eric Loken, Ken Nakayama, and Bradley Duchaine, "Human Face Recognition Ability Is Specific and Highly Heritable," *Proceedings of the National Academy of Sciences of the United States of America* 107, no. 11 (2010): 5238–5241; Nicholas G. Shakeshaft and Robert Plomin, "Genetic Specificity of Face Recognition," *Proceedings of the National Academy of Sciences of the United States of America* 112, no. 41 (2015): 12887–12892.

104 **individuals who are extraordinarily good:** Richard Russell, Brad Duchaine, and Ken Nakayama, "Super-Recognizers: People with Extraordinary Face Recognition Ability," *Psychonomic Bulletin and Review* 16, no. 2 (2009): 252–257.

105 **Both of these altruists were unaware:** Nicole Lyn Pesce, "New Yorker Gives Barefoot Homeless Woman Her Shoes on the Subway," *New York Daily News,* November 19, 2015, http://www.nydailynews.com/new-york/new-yorker-bare foot-homeless-woman-shoes-subway-article-1.2440107; Lucy Yang, "Good Samaritan Gives Shivering Man His Shirt, Hat on Subway," *ABC Eyewitness News,* January 10, 2016, http://abc7ny.com/society/exclusive-good-samaritan-who-gave -homeless-man-shirt-on-subway-speaks-out/1153750/.

105 **The 2016 World Giving Index estimates:** Charities Aid Foundation (CAF), "CAF World Giving Index 2016," https://www.cafonline.org/about-us/publications/2016-publications/world-giving-index-2016.

106 **The amount of money that Americans give:** Charity Navigator, "Giving Statistics," http://www.charitynavigator.org/index.cfm?bay=content.view&cpid=42#.VxV GaZMrIkg.

109 **one of Daniel Batson's studies of altruism:** C. Daniel Batson, Bruce D. Duncan, Paul Ackerman, Terese Buckley, and Kimberly Birch, "Is Empathic Emotion a Source of Altruistic Motivation?" *Journal of Personality and Social Psychology* 40, no. 2 (1981): 290–302.

110 **Two forces that biologists:** Abigail A. Marsh, "Neural, Cognitive, and Evolutionary Foundations of Human Altruism," *Wiley Interdisciplinary Reviews: Cognitive Science* 7, no. 1 (2016): 59–71.

112 **one frigid January afternoon in 1982:** Sue Anne Pressley Montes, "In a Moment of Horror, Rousing Acts of Courage," *Washington Post,* January 13, 2007; Blaine Harden, "Instant Hero," *Washington Post,* January 15, 1982; "Hero of Plane Crash Had Little Experience in the Hero Business," *Los Angeles Times/Washington Post* News Service, January 16, 1982.

114 **the moral equivalent of saving a drowning stranger:** D. Z. Levine, "When a Stranger Offers a Kidney: Ethical Issues in Living Organ Donation," *American Journal of Kidney Disease* 32, no. 4 (1998): 676–691.

114 **Before the 1990s, donating a kidney:** Reginald Y. Gohh, Paul E. Morrissey, Peter N. Madras, and Anthony P. Monaco, "Controversies in Organ Donation: The Altruistic Living Donor," *Nephrology Dialysis Transplantation* 16, no. 3 (2001): 619–621, DOI: https://doi.org/10.1093/ndt/16.3.619.

115 **These issues are rare, thankfully:** Anders Hartmann, Per Fauchald, Lars Westlie, Inge B. Brekke, and Hallvard Holdaas, "The Risks of Living Kidney Donation," *Nephrology Dialysis Transplantation* 18, no. 5 (2003): 871–873, DOI: https://doi.org /10.1093/ndt/gfg069.

115 **The odds of dying after tumbling out of a plane:** Jeremy Hus, "The Truth About Skydiving Risks," March 26, 2009, *LiveScience,* http://www.livescience.com/5350 -truth-skydiving-risks.html.

115 **if people who start out with above-average health:** Geir Mjøen, Stein Hallan, Anders Hartmann, Aksel Foss, Karsten Midtvedt, Ole Øyen, Anna Reisæter, Per Pfeffer, Trond Jenssen, Torbjørn Leivestad, Pål-Dag Line, Magnus Øvrehus, Dag Olav Dale, Hege Pihlstrøm, Ingar Holme, Friedo W. Dekker, and Hallvard Holdaas, "Long-Term Risks for Kidney Donors," *Kidney International* 86, no. 1 (2014): 162–167.

116 **"the first time in the history of medicine":** Francis D. Moore, "New Problems for Surgery," *Science* 144, no. 3617 (1964): 388–392, DOI: 10.1126/science .144.3617.388.

116 "have the Self primarily for their object": The misperception of human nature as uniform, and the problems this causes when it comes to transplantation decisions, was reinforced by David Levine, who wrote: "Despite the wide range of individual values, transplant centers often act as if there is a value consensus." Levine, "When a Stranger Offers a Kidney," 683.

117 Graef was not the first person: H. Harrison Sadler, Leslie Davison, Charles Carroll, and Samuel L. Kountz, "The Living, Genetically Unrelated, Kidney Donor," *Seminars in Psychiatry* 3, no. 1 (1971): 86–101; Levine, "When a Stranger Offers a Kidney."

119 This requirement, by the way: Levine, "When a Stranger Offers a Kidney."

120 So in early 2000, Dr. Gohh wrote up: Gohh et al., "Controversies in Organ Donation."

121 Nearly all transplant centers will now consent: Sadler et al., "The Living, Genetically Unrelated, Kidney Donor"; Aaron Spital, "Evolution of Attitudes at US Transplant Centers Toward Kidney Donation by Friends and Altruistic Strangers," *Transplantation* 69, no. 8 (2000): 1728–1731.

121 In 2009, I read "The Kindest Cut": Larissa MacFarquhar, "The Kindest Cut," *The New Yorker,* July 27, 2009.

121 My colleague David Rand: David G. Rand and Ziv G. Epstein, "Risking Your Life Without a Second Thought: Intuitive Decision-Making and Extreme Altruism," *PLoS One* 9, no. 10 (2014): e109687.

121 Altruistic kidney donors like Graef: Sadler et al., "The Living, Genetically Unrelated, Kidney Donor"; Lynn Stothers, William A. Gourlay, and Li Liu, "Attitudes and Predictive Factors for Live Kidney Donation: A Comparison of Live Kidney Donors Versus Nondonors," *Kidney International* 67, no. 3 (2005): 1105–1111. See also the TEDx Talk by altruistic kidney donor Ned Brooks, "What Makes a Person Decide to Donate His Kidney to a Stranger?" uploaded March 8, 2017, https://www.youtube.com/watch?v=nhht9kslq04. The unhesitating speed with which altruistic donors often make their decision is one source of the qualms of bioethicists regarding these donations, many of whom believe that truly informed consent must follow a period of careful deliberation (Levine, "When a Stranger Offers a Kidney").

Chapter 5: What Makes an Altruist?

126 Although most people believe that they are good: Aldert Vrij, Par Anders Granhag, and Stephen Porter, "Pitfalls and Opportunities in Nonverbal and Verbal Lie Detection," *Psychological Science in the Public Interest* 11, no. 3 (2010): 89–121.

126 Clues about others' emotions: Wen Zhou and Denise Chen, "Fear-Related Chemosignals Modulate Recognition of Fear in Ambiguous Facial Expressions," *Psychological Science* 20, no. 2 (2009): 177–183; Lilianne R. Mujica-Parodi, Helmut H. Strey, Blaise de B. Frederick, R. L. Savoy, David Cox, Yevgeny Botanov, Denis Tolkunov, Denis Rubin, and Jochen Weber, "Chemosensory Cues to Conspecific Emotional Stress Activate Amygdala in Humans," *PLoS One* 4, no. 7 (2009): e6415.

127 In 1978, Ekman and Friesen created: Paul Ekman and Wallace Friesen, *Pictures of Facial Affect* (Palo Alto, CA: Consulting Psychologists, 1976).

127 Ekman and Friesen determined that for a face: Paul Ekman, Wallace V. Friesen, and Joseph C. Hager, Facial Action Coding System: A Technique for the Measurement of Facial Movement" (Palo Alto, CA: Consulting Psychologists, 1978).

127 **Human eyes are ideally designed:** Hiromi Kobayashi and Shiro Kohshima, "Unique Morphology of the Human Eye," *Nature* 387 (1997): 767–768.

128 **Ekman has observed that the expressive muscles:** David Matsumoto and Paul Ekman, "Facial Expression Analysis," *Scholarpedia* 3, no. 5 (2008): 4237, http://www.scholarpedia.org/article/Facial_expression_analysis.

131 **Our speed has slowed but is still blisteringly fast:** Jitendra Malik, "Human Visual System" (lecture), University of California at Berkeley, January 27, 2004, https://people.eecs.berkeley.edu/~malik/cs294/lecture2-RW.pdf.

131 **Findings presented in a 2016 article:** Constantino Méndez-Bértolo, Stephan Moratti, Rafael Toledano, Fernando Lopez-Sosa, Roberto Martínez-Alvarez, Yee H. Mah, Patrik Vuilleumier, Antonio Gil-Nagel, and Bryan A. Strange, "A Fast Pathway for Fear in Human Amygdala," *Nature Neuroscience* 19, no. 8 (2016): 1041–1049, DOI: 10.1038/nn.4324.

132 **They found that the amygdala:** Paul J. Whalen, Jerome Kagan, Robert G. Cook, F. Caroline Davis, Hackjin Kim, Sara Polis, Donald G. McLaren, Leah H. Somerville, Ashly A. McLean, Jeffrey S. Maxwell, and Tom Johnstone, "Human Amygdala Responsivity to Masked Fearful Eye Whites," *Science* 306, no. 5704 (2004): 2061.

133 **Birds can recognize the alarm calls:** Hugo J. Rainey, Klaus Zuberbühler, and Peter J. Slater, "Hornbills Can Distinguish Between Primate Alarm Calls," *Proceedings of the Royal Society: Biological Sciences* 271, no. 1540 (2004): 755–759.

133 **Many have argued or assumed:** Marsh, "Understanding Amygdala Responsiveness to Fearful Expressions."

134 **the fearlike facial expressions of other primates:** A. Parr and Bridget M. Waller, "Understanding Chimpanzee Facial Expression: Insights into the Evolution of Communication," *Social Cognitive and Affective Neuroscience* 1, no. 3 (2006): 221–228, DOI: 10.1093/scan/nsl031.

134 **Angry faces actually generate even less:** This is true in both humans and monkeys; see Fusar-Poli et al., "Functional Atlas of Emotional Faces Processing"; Ning Liu, Fadila Hadj-Bouziane, Katherine B. Jones, Janita N. Turchi, Bruno B. Averbeck, and Leslie G. Ungerleider, "Oxytocin Modulates fMRI Responses to Facial Expression in Macaques," *Proceedings of the National Academy of Sciences of the United States of America* 112, no. 24 (2015): E3123–E3130.

136 **It's this simulation that causes faint whispers:** Yvonne Rothemund, Silvio Ziegler, Christiane Hermann, Sabine M. Gruesser, Jens Foell, Christopher J. Patrick, and Herta Flor, "Fear Conditioning in Psychopaths: Event-Related Potentials and Peripheral Measures," *Biological Psychology* 90 (2012): 50–59; Antoine Bechara, Hanna Damasio, Antonio R. Damasio, and Gregory P. Lee, "Different Contributions of the Human Amygdala and Ventromedial Prefrontal Cortex to Decision-Making," *Journal of Neuroscience* 19, no. 13 (1999): 5473–5481.

136 **The uncanny overlap in the regions:** Claus Lamm, Jean Decety, and Tania Singer, "Meta-Analytic Evidence for Common and Distinct Neural Networks Associated with Directly Experienced Pain and Empathy for Pain," *Neuroimage* 54, no. 3 (2011): 2492–2502. Perhaps the most compelling evidence for the insula response to others' pain representing an empathic phenomenon comes from a study that found that participants who were given a placebo experienced both reductions in pain and reduced insula responses to others' pain—and that both of these effects could be eliminated by administering a drug called naltrexone,

which blocks brain receptors to opioids, the neurotransmitters involved in pain responding; see Markus Rütgen, Eva-Maria Seidel, Giorgia Silani, Igor Riečanský, Allan Hummer, Christian Windischberger, Predrag Petrovic, and Claus Lamm, "Placebo Analgesia and Its Opioidergic Regulation Suggest That Empathy for Pain Is Grounded in Self Pain," *Proceedings of the National Academy of Sciences of the United States of America* (2015): E5638–E5646.

137 **a clever brain imaging study:** Grit Hein, Giorgia Silani, Kerstin Preuschoff, C. Daniel Batson, and Tania Singer, "Neural Responses to Ingroup and Outgroup Members' Suffering Predict Individual Differences in Costly Helping," *Neuron* 68, no. 1 (2010): 149–160.

139 **One recent study investigating the acoustic properties:** Luc H. Arnal, Adeen Flinker, Andreas Kleinschmidt, Anne-Lise Giraud, and David Poeppel, "Human Screams Occupy a Privileged Niche in the Communication Soundscape," *Current Biology* 25, no. 15 (2015): 2051–2056.

139 **Studies of patients with amygdala lesions:** Reiner Sprengelmeyer, Andrew W. Young, Ulrike Schroeder, Peter G. Grossenbacher, Jens Federlein, Thomas Buttner, and Horst Przuntek, "Knowing No Fear," *Proceedings of the Royal Society of London B: Biological Sciences* 266, no. 1437 (1999): 2451–2456; Nathalie Gosselin, Isabelle Peretz, Erica Johnsen, and Ralph Adolphs, "Amygdala Damage Impairs Emotion Recognition from Music," *Neuropsychologia* 45, no. 2 (2007): 236–244.

139 **the amygdala is also important for identifying behaviors:** Abigail A. Marsh and Elise M. Cardinale, "Psychopathy and Fear: Specific Impairments in Judging Behaviors That Frighten Others," *Emotion* 12, no. 5 (2012): 892–898; Abigail A. Marsh and Elise M. Cardinale, "When Psychopathy Impairs Moral Judgments: Neural Responses During Judgments About Causing Fear," *Social Cognitive and Affective Neuroscience* 9 (2014): 3–11.

140 **one study of adult psychopaths:** Patricia L. Lockwood, Catherine L. Sebastian, Eamon J. McCrory, Zoe H. Hyde, Xiaosi Gu, Stéphane A. De Brito, and Essi Viding, "Association of Callous Traits with Reduced Neural Response to Others' Pain in Children with Conduct Problems," *Current Biology* 23, no. 10 (2013): 901–905; Abigail A. Marsh, Elizabeth C. Finger, Katherine A. Fowler, Christopher J. Adalio, Ilana T. N. Jurkowitz, Julia C. Schechter, Daniel S. Pine, Jean Decety, and Robert James R. Blair, "Empathic Responsiveness in Amygdala and Anterior Cingulate Cortex in Youths with Psychopathic Traits," *Journal of Child Psychology and Psychiatry* 54, no. 8 (2013): 900–910; Jean Decety, Laurie R. Skelly, and Kent A. Kiehl, "Brain Response to Empathy-Eliciting Scenarios Involving Pain in Incarcerated Individuals with Psychopathy," *JAMA Psychiatry* 70, no. 6 (2013): 638–645.

147 **Belay recalled that her doctor:** "1-800-Give-Us-Your-Kidney," Conscious Good, 2016, https://www.consciousgood.com/1-800-give-us-your-kidney/.

150 **Half a cubic centimeter or so of flesh:** Abigail A. Marsh, Sarah A. Stoycos, Kristin M. Brethel-Haurwitz, Paul Robinson, John W. VanMeter, and Elise M. Cardinale, "Neural and Cognitive Characteristics of Extraordinary Altruists," *Proceedings of the National Academy of Sciences of the United States of America* 111, no. 42 (2014): 15036–15041.

150 **they show heightened amygdala responses:** Murray B. Stein, Alan N. Simmons, Justin S. Feinstein, B. S. Martin P. Paulus, "Increased Amygdala and Insula Activation During Emotion Processing in Anxiety-Prone Subjects," *American Journal of Psychiatry* 164, no. 2 (2007): 318–327; K. Luan Phan, Daniel A. Fitzgerald,

Pradeep J. Nathan, and Manuel E. Tancer, "Association Between Amygdala Hyperactivity to Harsh Faces and Severity of Social Anxiety in Generalized Social Phobia," *Biological Psychiatry* 59, no. 5 (2006): 424–429; Murray B. Stein, Philippe R. Goldin, Jitender Sareen, Lisa T. Eyler, and Gregory G. Brown, "Increased Amygdala Activation to Angry and Contemptuous Faces in Generalized Social Phobia," *Archives of General Psychiatry* 59, no. 11 (2002): 1027–1034.

153 **during an interview with Ted Koppel:** Blaine Harden, "Instant Hero," *Washington Post,* January 15, 1982.

154 **"Writers of sketches, in a friendly desire":** Clara Barton, *The Story of My Childhood* (New York: Baker & Taylor Co., 1907), 15.

155 **echo the words of Lenny Skutnik:** Marlyn Schwartz, "Has Fame Spoiled Lenny Skutnik?" *New York Times* News Service, March 24, 1982, https://news.google.com/newspapers?nid=1734&dat=19820324&id=ZVQcAAAAIBAJ&sjid=elIEAAAAIBAJ&pg=5381,8431318&hl=en.

Chapter 6: The Milk of Human Kindness

158 **only one of every 1,000 loggerhead babies:** Nat B. Frazer, "Survival from Egg to Adulthood in a Declining Population of Loggerhead Turtles, Caretta Caretta," *Herpetologica* (1986): 47–55.

161 **"absurd in the highest possible degree":** Charles Darwin, *The Origin of Species* (New York: Collier & Son, 1909), 190.

164 **turtles predate even the dinosaurs:** Rosensteil School of Marine and Atmospheric Science, University of Miami, "Sea Turtle History," http://www.rsmas.miami.edu/outreach/explore-and-discover/sea-turtles/history/; "Turtles: History and Fossil Record," http://science.jrank.org/pages/7044/Turtles-History-fossil-record.html.

164 **hamsterlike creatures called cynodonts:** Olav T. Oftedal, "The Mammary Gland and Its Origin During Synapsid Evolution," *Journal of Mammary Gland Biology and Neoplasia* 7, no. 3 (2002): 225–252.

165 **This elixir is directly responsible:** Caroline M. Pond, "The Significance of Lactation in the Evolution of Mammals," *Evolution* (1977): 177–199.

167 **the capacity for love and caring of all kinds**: C. Daniel Batson, "The Naked Emperor: Seeking a More Plausible Genetic Basis for Psychological Altruism," *Economics and Philosophy* 26, no. 2 (2010): 149–164.

167 **"a turning point in the evolution":** Irenäus Eibl-Eibesfeldt, *Love and Hate: The Natural History of Behavior Patterns* (Chicago: Aldine, 1996), xi.

168 **this bundling is in perfect accordance:** Peter H. Klopfer, "Origins of Parental Care," in *Parental Care in Mammals,* edited by David J. Gubernick and Peter H. Klopfer (New York: Plenum, 1981) 1–12.

170 **the ewe will reserve all of her nurturing and milk:** Keith M. Kendrick, Ana P. C. Da Costa, Kevin D. Broad, Satoshi Ohkura, Rosalinda Guevara, Frederic Lévy, and E. Barry Keverne, "Neural Control of Maternal Behaviour and Olfactory Recognition of Offspring," *Brain Research Bulletin* 44, no. 4 (1997): 383–395; E. Barry Keverne and Keith M. Kendrick, "Oxytocin Facilitation of Maternal Behavior in Sheep," *Annals of the New York Academy of Sciences* 652 (1992): 83–101; Larry J. Young and Thomas R. Insel, "Hormones and Parental Behavior," in *Behavioral Endocrinology,* 2nd ed., edited by Jill B. Becker, S. Marc Breedlove, David Crews, and Margaret M. McCarthy (Cambridge, MA: MIT Press), 331–366.

170 **unearthed a long-forgotten 1968 study:** William E. Wilsoncroft, "Babies by Bar-Press: Maternal Behavior in the Rat," *Behavior Research Methods and Instrumentation* 1 (1968): 229–230; Stephanie D. Preston, "The Origins of Altruism in Offspring Care," *Psychological Bulletin* 139, no. 6 (2013): 1305–1341.

172 **It's a behavior that is at least three times more likely:** Sarah Blaffer Hrdy, *Mothers and Others: The Evolutionary Origins of Mutual Understanding* (Cambridge, MA: Harvard University Press, 2009).

172 **Among the many other mammals:** Marianne L. Riedman, "The Evolution of Alloparental Care and Adoption in Mammals and Birds," *Quarterly Review of Biology* 57 (1982): 405–435.

173 **One spectacular demonstration:** Marc Lacey, "5 Little Oryxes and the Big Bad Lioness of Kenya," *New York Times,* October 12, 2002; Anthony Yap, "Kamunyak, the Blessed One: The Lioness Who Adopts Oryx Calves," Phantom Maelstrom, November 29, 2011, http://phantommaelstrom.blogspot.com/2011/11/kamunyak-blessed-one-lioness-who-adopts.html; "The Lioness and the Oryx," *BBC News,* January 7, 2002, http://news.bbc.co.uk/2/hi/africa/1746828.stm.

173 **Yet still she cared for the little calf:** "The Lioness and the Oryx," *Nat Geo Wild,* http://channel.nationalgeographic.com/wild/unlikely-animal-friends/videos/the-lioness-and-the-oryx/.

174 **Another lioness in Uganda:** Emma Reynolds, "Extraordinary Moment Wounded Lioness Shows Softer Side by Adopting Baby Antelope (Perhaps She Was Feeling Guilty After Killing Its Mother)," *The Daily Mail,* October 8, 2012.

174 **images of the baby trying to suckle:** Paul Steyn, "Cat Watch: Baby Baboon's Frightening Encounter with Lions Ends with a Heroic Twist," *National Geographic,* April 3, 2014, http://voices.nationalgeographic.com/2014/04/03/baby-baboons-dramatic-encounter-with-lions-ends-with-a-heroic-twist.

175 **a grizzled ten-year-old female chihuahua:** "Ai Chihuahua! Dog Adopts 4 Baby Squirrels," CowboysZone, September 8, 2007, http://cowboyszone.com/threads/ai-chihuahua-dog-adopts-4-baby-squirrels.94635/; "Chihuahua Mothers Abandoned Baby Squirrels," For the Love of the Dog Blog, September 8, 2007, http://fortheloveofthedogblog.com/news-updates/chihuahua-mothers-abandoned-baby-squirrels.

175 **"resident nursery companion" at the Cincinnati Zoo:** Kelli Bender, "Blakely Plays the Role of Dog Dad to Ohio Zoo's Rejected Baby Takin," *People,* August 18, 2015, http://site.people.com/pets/blakely-plays-the-role-of-dog-dad-to-ohio-zoos-rejected-baby-takin-video/.

175 **a golden retriever named Izzy:** Mike Celizic, "Tigers Say 'Bye Mom' to Dog That Raised Them," *Today,* June 25, 2009, http://www.today.com/id/31541834/ns/today-today_pets/t/tigers-say-bye-mom-dog-raised-them/#.V-SKiJMrLkI.

176 **From tiny tamarins and marmosets to siamangs:** Karen Isler and Carel P. van Schaik, "Allomaternal Care, Life History, and Brain Size Evolution in Mammals," *Journal of Human Evolution* 63, no. 1 (2012): 52–63.

176 **real allomothering superstars are humans:** Hrdy, *Mothers and Others.*

177 **Humans in modern cultures:** Courtney L. Meehan and Alyssa N. Crittenden, *Childhood: Origins, Evolution, and Implications* (Albuquerque: University of New Mexico Press, 2016).

177 **tracing back to the psychiatrist John Bowlby:** John Bowlby, *A Secure Base: Parent-Child Attachment and Healthy Human Development* (New York: Basic

Books, 2008); Mary D. Ainsworth, "Infant-Mother Attachment," *American Psychologist* 34, no. 10 (1979): 932–937.

177 whether child care for working mothers: D'Vera Cohn and Andrea Caumont, "7 Key Findings About Stay-at-Home Moms," Pew Research Center, April 8, 2014, http://www.pewresearch.org/fact-tank/2014/04/08/7-key-findings-about-stay-at-home-moms/.

178 "Children do best in societies": Quoted in Hrdy, *Mothers and Others,* 103.

178 mothers in some foraging cultures: Ibid., 100.

179 Inadequate social support is a top risk factor: Emma Robertson, Sherry Grace, Tamara Wallington, and Donna E. Stewart, "Antenatal Risk Factors for Postpartum Depression: A Synthesis of Recent Literature," *General Hospital Psychiatry* 26, no. 4 (2004): 289–295.

179 the famous "napalm girl" photograph: Gendy Alimurung, "Nick Ut's *Napalm Girl* Helped End the Vietnam War. Today in LA, He's Still Shooting," *LA Weekly,* July 17, 2014, http://www.laweekly.com/news/nick-uts-napalm-girl-helped-end-the-vietnam-war-today-in-la-hes-still-shooting-4861747.

179 the awful, heartrending image of Aylan Kurdi: Roy Greenslade, "So Aylan Kurdi's Picture Did Make a Difference to the Refugee Debate," *The Guardian,* September 4, 2015; Jessica Elgot, "Charity Behind Migrant-Rescue Boats Sees 15-Fold Rise in Donations in 24 Hours," *The Guardian,* September 3, 2015; Paul Slovic, Daniel Västfjäll, Arvid Erlandsson, and Robin Gregory, "Iconic Photographs and the Ebb and Flow of Empathic Response to Humanitarian Disasters," Proceedings of the National Academy of Sciences of the United States of America 114, no. 4 (2017): 640–644, DOI: 10.1073/pnas.1613977114.

180 what ethologists call "key stimuli": Leslie A. Zebrowitz, *Reading Faces: The Window to the Soul?* (Boulder, CO: Westview Press, 1997), 68; Eibl-Eibesfeldt, *Love and Hate.*

180 their brains and the tops of their skulls: Doug Jones, C. Loring Brace, William Jankowiak, Kevin N. Laland, Lisa E. Musselman, Judith H. Langlois, Lori A. Roggman, Daniel Pérusse, Barbara Schweder, and Donald Symons, "Sexual Selection, Physical Attractiveness, and Facial Neoteny: Cross-Cultural Evidence and Implications," *Current Anthropology* 36 (1995): 723–748.

180 The resulting babyish proportions: Konrad Lorenz, "Die angeborenen Formen moeglicher Erfahrung," *Zeitschrift fur Tierpsychologie* 5 (1943): 235–409.

180 after adult men and women simply view: Gary D. Sherman, Jonathan Haidt, and James A. Coan, "Viewing Cute Images Increases Behavioral Carefulness," *Emotion* 9, no. 2 (2009): 282–286.

181 Subjects who were less psychopathic: Jennifer L. Hammer and Abigail A. Marsh, "Why Do Fearful Facial Expressions Elicit Behavioral Approach? Evidence from a Combined Approach-Avoidance Implicit Association Test," *Emotion* 15 (2015): 223–231.

181 a cute, appealing, babyish appearance: Zebrowitz, *Reading Faces.*

181 These effects are not simply due: Caroline F. Keating, David W. Randall, Timothy Kendrick, and Katharine A. Gutshall, "Do Babyfaced Adults Receive More Help? The (Cross-Cultural) Case of the Lost Résumé," *Journal of Nonverbal Behavior* 27, no. 2 (2003): 89–109; David A. Lishner, Luis V. Oceja, E. L. Stocks, and Kirstin Zaspel, "The Effect of Infant-Like Characteristics on Empathic Concern for Adults in Need," *Motivation and Emotion* 32 (2008): 270–277.

182 **Over half of all American households:** For a useful summary of academic efforts to explain pet-keeping, with various levels of success, see Melissa Hogenboom, "Why Do We Love Our Pets So Much?" *BBC Earth,* May 29, 2015, http://www.bbc.com/earth/story/20150530-why-do-we-love-our-pets-so-much.

182 **one of the causes of rising pet ownership:** For articles delving into this possibility, see Robert A. Ferdman, "Modern Family: Americans Are Having Dogs Instead of Babies," *Quartz,* April 10, 2014, http://qz.com/197416/americans-are-having-dogs-instead-of-babies/; Karen E. Bender, "Dogs: The Best Kids You Could Ask For," *The Atlantic,* August 22, 2014; Jordan Weissmann, "Why America's Falling Birth Rate Is Sensational News for the Pet Industry," *The Atlantic,* May 20, 2013.

183 **a Russian motorist risked his life:** Jenny Starrs, "Harrowing Footage Shows Motorists Dodging Kitten on Busy Russian Highway—Until One Man Stops," *Washington Post,* September 16, 2016.

184 **allomothering provides the basis for altruism:** Judith Maria Burkart, O. Allon, Federica Amici, Claudia Fichtel, Christia Finkenwirth, Adolf Heschl, J. Huber, Karin Isler, Z. K. Kosonen, E. Martins, Ellen J. M. Meulman, R. Richiger, K. Rueth, B. Spillmann, S. Wiesendanger, and Carel P. van Schaik, "The Evolutionary Origin of Human Hyper-Cooperation," *Nature Communications* 5 (2014): 4747.

186 **Fearful eyes are wide and large:** Roger Segelken, "Survey Explains Why Some Animals Have Smaller Eyes: Lifestyle Matters More Than Size, Cornell Biologists Say," *Cornell Chronicle,* August 6, 2004, http://www.news.cornell.edu/stories/2004/08/why-some-animals-have-smaller-eyes-lifestyle-matters.

186 **adopting a fearful expression causes a face:** Abigail A. Marsh, Reginald B. Adams Jr., and Robert E. Kleck, "Why Do Fear and Anger Look the Way They Do? Form and Social Function in Facial Expressions," *Personality and Social Psychology Bulletin* 31, no. 1 (2005): 73–86.

187 **some of our nearest primate relatives:** Hillary Anger Elfenbein and Nalini Ambady, "On the Universality and Cultural Specificity of Emotion Recognition: A Meta-Analysis," *Psychological Bulletin* 128, no. 2 (2002): 203–235; Signe Preuschoft, "Primate Faces and Facial Expressions." *Social Research* 67, no. 1 (2000): 245–271.

188 **key traits of the one creature that social mammals:** Rudolf Schenkel, "Submission: Its Features and Function in the Wolf and Dog," *American Zoologist* 7 (1967): 319–329.

188 **"The plaintive voice of misery":** Adam Smith, *The Theory of Moral Sentiments* (1853), 48.

189 **the entry point into the parental care system:** Stephanie D. Preston, "The Origins of Altruism in Offspring Care," *Psychological Bulletin* 139, no. 6 (2013): 1305–1341.

189 **regardless of whether the cues:** Leslie A. Zebrowitz, Victor X. Luevano, P. Matthew Bronstad, and Itzhak Aharon, "Neural Activation to Babyfaced Men Matches Activation to Babies," *Social Neuroscience* 4, no. 1 (2009): 1–10; Chris Baeken, Rudi De Raedt, Nick F. Ramsey, Peter Van Schuerbeek, Dora Hermes, Axel Bossuyt, Lemke Leyman, Marie-Anne Vanderhasselt, Johan De Mey, and Robert Luypaert, "Amygdala Responses to Positively and Negatively Valenced Baby Faces in Healthy Female Volunteers: Influences of Individual Differences in Harm Avoidance," *Brain Research* 1296 (2009): 94–103. Note that, in new mothers, amygdala activation may be strongest specifically to images of one's

own baby; see S. Ranote, Rebecca Elliott, Kathryn M. Abel, Rachel Mitchell, John Francis William Deakin, and L. Appleby, "The Neural Basis of Maternal Responsiveness to Infants: An fMRI Study," *Neuroreport* 15, no. 11 (2004): 1825–1829.

189 **listening to them results in more activation:** Amygdala responsiveness to cries varies impressively as a function of a number of variables, including gender, parental status, personality, and exactly whose infant is crying, but some level of amygdala responsiveness to infant cries is a constant across nearly every study on the topic; see Kerstin Sander, Yvonne Frome, and Henning Scheich, "fMRI Activations of Amygdala, Cingulate Cortex, and Auditory Cortex by Infant Laughing and Crying," *Human Brain Mapping* 28, no. 10 (2007): 1007–1022; Erich Seifritz, Fabrizio Esposito, John G. Neuhoff, Andreas Lüthi, Henrietta Mustovic, Gerhard Dammann, Ulrich von Bardeleben, Ernst W. Radue, Sossio Cirillo, Gioacchino Tedeschi, and Francesco Di Salle, "Differential Sex-Independent Amygdala Response to Infant Crying and Laughing in Parents Versus Nonparents," *Biological Psychiatry* 54, no. 12 (2003): 1367–1375; Isabella Mutschler, Tonio Ball, Ursula Kirmse, Birgit Wieckhorst, Michael Pluess, Markus Klarhöfer, Andrea H. Meyer, Frank H. Wilhelm, and Erich Seifritz, "The Role of the Subgenual Anterior Cingulate Cortex and Amygdala in Environmental Sensitivity to Infant Crying," *PLoS One* 11, no. 8 (2016): e0161181.

191 **differs from oxytocin by only one amino acid:** Valery Grinevich, H. Sophie Knobloch-Bollmann, Marina Eliava, Marta Busnelli, and Bice Chini, "Assembling the Puzzle: Pathways of Oxytocin Signaling in the Brain," *Biological Psychiatry* 79, no. 3 (2015): 155–164. Incidentally, the limited parental care that sea turtles supply is directly supported by vasotocin. When a mother turtle begins to drag herself up onto the beach in preparation for nesting, the amount of vasotocin produced in her hypothalamus begins to climb. As she begins to dig a hole in the sand with her flippers, it increases still further, then spikes dramatically when the nest nears completion. At the moment she deposits her first egg in the nest, vasotocin levels are 1,500 times greater than when they started rising. Production of vasotocin starts to drop shortly thereafter, remaining just detectable as she covers her nest up with sand. By the time she slips back into the sea, vasotocin production has dwindled away to almost nothing. Does this surge in vasotocin feel like anything to a loggerhead? Does some glimmer of maternal concern pass through her craggy head as she tamps sand over her nest? It is impossible to say, although the fact that sea turtles weep briny tears as they lay their ill-fated eggs has caused some to speculate. See Robert A. Figler, Duncan S. MacKenzie, David W. Owens, Paul Licht, and Max S. Amoss, "Increased Levels of Arginine Vasotocin and Neurophysin During Nesting in Sea Turtles," *General and Comparative Endocrinology* 73, no. 2 (1989): 223–232.

191 **the pituitary releases the oxytocin:** C. Sue Carter and Margaret Altemus, "Integrative Functions of Lactational Hormones in Social Behavior and Stress Management," *Annals of the New York Academy of Sciences* 807 (1997): 164–74.

193 **injected oxytocin into female rats' brains:** Cort A. Pedersen, John A. Ascher, Yvonne L. Monroe, and Arthur J. Prange Jr., "Oxytocin Induces Maternal Behavior in Virgin Female Rats," *Science* 216, no. 4546 (1982): 648–650.

194 **Oxytocin has been shown to induce:** I should emphasize that any complex behavior involves a complex array of processes within the brain, and maternal care is no different. Given this, the widespread agreement that oxytocin's effects on subcortical structures like the amygdala are the most essential processes supporting neural care is impressive. For reviews on the effects of oxytocin on maternal behavior, see Keith M. Kendrick, "Neural Control of Maternal Behaviour and Olfactory Recognition of Offspring," *Brain Research Bulletin* 44, no. 4 (1997): 383–395; Thomas R. Insel and Lawrence E. Shapiro, "Oxytocin Receptors and Maternal Behavior," *Annals of the New York Academy of Sciences* 652 (1992): 122–141; C. Sue Carter, "Neuroendocrine Perspectives on Social Attachment and Love," *Psychoneuroendocrinology* 23, no. 8 (1998): 779–818; Gareth Leng, Simone L. Meddle, and Alison J. Douglas, "Oxytocin and the Maternal Brain," *Current Opinion in Pharmacology* 8, no. 6 (2008): 731–734.

194 the amygdala is a central locus: Oliver J. Bosch and Inga D. Neumann, "Both Oxytocin and Vasopressin Are Mediators of Maternal Care and Aggression in Rodents: From Central Release to Sites of Action," *Hormones and Behavior* 61, no. 3 (2012): 293–303; Thomas R. Insel, "The Challenge of Translation in Social Neuroscience: A Review of Oxytocin, Vasopressin, and Affiliative Behavior," *Neuron* 65, no. 6 (2010): 768–779.

195 News articles suggested that car dealerships: Stefan Lovgren, "'Trust' Hormone's Smell Helps Us Hand over Cash, Study Says," *National Geographic News,* June 1, 2005, http://news.nationalgeographic.com/news/2005/06/0601_050601_trustpotion.html.

196 a few squirts of oxytocin up each nostril: Abigail A. Marsh, Henry H. Yu, Daniel S. Pine, Elena K. Gorodetsky, David Goldman, and R. J. R. Blair, "The Influence of Oxytocin Administration on Responses to Infant Faces and Potential Moderation by OXTR Genotype," *Psychopharmacology* 224, no. 4 (2012): 469–476.

196 several other researchers have since reported: Abigail A. Marsh, Henry H. Yu, Daniel S. Pine, and R. J. Blair, "Oxytocin Improves Specific Recognition of Positive Facial Expressions," *Psychopharmacology (Berl)* 209, no. 3 (2010): 225–232; Meytal Fischer-Shofty, Simone G. Shamay-Tsoory, Hagai Harari, and Yechiel Levkovitz, "The Effect of Intranasal Administration of Oxytocin on Fear Recognition," *Neuropsychologia* 48, no. 1 (2010): 179–184; Meytal Fischer-Shofty, Simone G. Shamay-Tsoory, and Yechiel Levkovitz, "Characterization of the Effects of Oxytocin on Fear Recognition in Patients with Schizophrenia and in Healthy Controls," *Frontiers in Neuroscience* 7 (2013): 127; Alexander Lischke, Christoph Berger, Kristin Prehn, Markus Heinrichs, Sabine C. Herpertz, and Gregor Domes, "Intranasal Oxytocin Enhances Emotion Recognition from Dynamic Facial Expressions and Leaves Eye-Gaze Unaffected," *Psychoneuroendocrinology* 37, no. 4 (2012): 475–481.

197 the findings reported in a 2016 study: Daniele Viviani, Alexandre Charlet, Erwin van den Burg, Camille Robinet, Nicolas Hurni, Marios Abatis, Fulvio Magara, and Ron Stoop, "Oxytocin Selectively Gates Fear Responses Through Distinct Outputs from the Central Amygdala," *Science* 333, no. 6038 (2011): 104–107.

197 Their courage seems to result: Oliver Bosch, Simone L. Meddle, Daniela I. Beiderbeck, Alison J. Douglas, and Inga D. Neumann, "Brain Oxytocin Correlates with Maternal Aggression: Link to Anxiety," *Journal of Neuroscience* 25, no. 29 (2005): 6807–6815; Oliver J. Bosch, "Maternal Nurturing Is Dependent on Her Innate Anxiety: The Behavioral Roles of Brain Oxytocin and Vasopressin," *Hormones and Behavior* 59, no. 2 (2011): 202–212.

198 These neurons may signal other cells: For evidence that the amygdala may mediate what would otherwise be a fearful response to infants, see Alison S. Fleming,

Frank Vaccarino, and Carola Luebke, "Amygdaloid Inhibition of Maternal Behavior in the Nulliparous Female Rat," *Physiology and Behavior* 25, no. 5 (1980): 731–743. This very recent paper on rats provides strong support for this hypothesis: Elizabeth Rickenbacher, Rosemarie E. Perry, Regina M. Sullivan, and Marta A. Moita. "Freezing Suppression By Oxytocin in Central Amygdala Allows Alternate Defensive Behaviours and Mother-Pup Interactions." *Elife* 6 (2017), e24080.

198 **Dysfunction throughout the amygdala:** Mark R. Dadds, Caroline Moul, Avril Cauchi, Carol Dobson-Stone, David J. Hawes, John Brennan, and Richard E. Ebstein, "Methylation of the Oxytocin Receptor Gene and Oxytocin Blood Levels in the Development of Psychopathy," *Development and Psychopathology* (2014): 33–40; Mark R. Dadds, Caroline Moul, Avril Cauchi, Carol Dobson-Stone, David J. Hawes, John Brennan, Ruth Urwin, and Richard E. Ebstein, "Polymorphisms in the Oxytocin Receptor Gene Are Associated with the Development of Psychopathy," *Development and Psychopathology* (2013): 1–11; Joseph H. Beitchman, Clement C. Zai, Katherine Muir, Laura Berall, Behdin Nowrouzi-Kia, Esther Choi, and James L. Kennedy, "Childhood Aggression, Callous-Unemotional Traits, and Oxytocin Genes," *European Child and Adolescent Psychiatry* 21 (2012): 125–132.

Chapter 7: Can We Be Better?

203 **Gallup polls of thousands of people:** Charities Aid Foundation (CAF), "CAF World Giving Index 2016," https://www.cafonline.org/about-us/publications/2016-publications/caf-world-giving-index-2016.

203 **Americans donate hundreds of billions of dollars:** Corporation for National & Community Service, "Volunteering and Civic Life in America," https://www.nationalservice.gov/vcla.

204 **Americans donate over 13 million units of their blood:** AABB, "Blood FAQ," http://www.aabb.org/tm/Pages/bloodfaq.aspx.

204 **Thousands more Americans undergo:** US Department of Health and Human Services, Health Resources and Services Administration, "General FAQ," https://bloodcell.transplant.hrsa.gov/about/general_faqs/.

204 **respondents reported having treated:** National Wildlife Rehabilitators Association (NWRA), "Facts About NWRA," http://www.nwrawildlife.org/?page=Facts.

204 **international wildlife rehabilitation groups:** "International Wildlife Rehabilitators," http://wildlife.rescueshelter.com/international.

205 **Per capita, Americans donated:** Charity Navigator, "Giving Statistics," http://www.charitynavigator.org/index.cfm/bay/content.view/cpid/42; Philanthropy Roundtable, "Statistics," http://www.philanthropyroundtable.org/almanac/statistics/.

205 **Globally, blood donations are also increasing:** World Health Organization (WHO), "Blood Safety and Availability," July 2016, http://www.who.int/mediacentre/factsheets/fs279/en/.

206 **over three times as many people received marrow:** Figure 1 from Dennis Confer and Pam Robinett, "The US National Marrow Donor Program Role in Unrelated Donor Hematopoietic Cell Transplantation," *Bone Marrow Transplantation* 42, suppl. 1 (2008): S3–S5.

206 **Bureau of Labor Statistics (BLS) estimates:** Andy Kiersz, "Volunteering in America Is at Its Lowest Level in at Least a Decade," *Business Insider*, February 25, 2014, http://www.businessinsider.com/bls-volunteering-chart-2014-2; BLS, "Volunteering in the United States, 2015," February 25, 2016, https://www.bls.gov/news.release/volun.nr0.htm.

206 **the trend may instead reflect forces that affect volunteering:** "Americans with No Religious Affiliation," from Pew Research Center, *2014 Religious Landscape Study*, http://static6.businessinsider.com/image/55526680ecad04ac07fbd880-1200-900/godless-millennials.png; Corporation for National & Community Service, "National: Trends and Highlights Overview," https://www.nationalservice.gov/vcla/national.

207 **Steven Pinker has provided convincing evidence:** Steven Pinker, *The Better Angels of Our Natures: Why Violence Has Declined* (New York: Viking, 2011).

207 **In Europe, the homicide rate today:** Steven Pinker, "A History of Violence: Edge Master Class 2011," September 27, 2011, https://www.edge.org/conversation/steven_pinker-a-history-of-violence-edge-master-class-2011.

208 **Mauritania's abolition of slavery in 1980:** Ibid.

208 **only a minority of polled Americans:** "The Death Penalty, Nearing Its End" (editorial), *New York Times*, October 24, 2016.

208 **no war zones anywhere in the Western Hemisphere:** Greg Myre, "How Castro's Rise and Death Bookend 60 Years of Latin American Wars," NPR, September 27, 2016, http://www.npr.org/sections/parallels/2016/09/27/495522306/guess-what-as-of-today-the-western-hemisphere-has-no-wars.

208 **a majority of Americans have reported:** Pew Research Center, "Public Perception of Crime Rate at Odds with Reality," April 16, 2015, http://www.pewresearch.org/fact-tank/2015/04/17/despite-lower-crime-rates-support-for-gun-rights-increases/ft_15-04-01_guns_crimerate/.

208 **Although youth violence and delinquency:** Will Dahlgreen, "British Public Unaware of Improvement in Youth Behaviour," March 3, 2015, YouGovUK, https://yougov.co.uk/news/2015/03/03/british-public-unaware-revolution-youth-behaviour/.

209 **especially biased toward focusing on bad things:** Roy F. Baumeister, Ellen Bratslavsky, Catrin Finkenauer, and Kathleen D. Vohs, "Bad Is Stronger Than Good," *Review of General Psychology* 5, no. 4 (2001): 323.

209 **we generally pay more attention to bad events:** For compelling examples of this phenomenon, see Joop Van der Pligt and J. Richard Eiser, "Negativity and Descriptive Extremity in Impression Formation," *European Journal of Social Psychology* 10, no. 4 (1980): 415–419; Nicole C. Baltazar, Kristin Shutts, and Katherine D. Kinzler, "Children Show Heightened Memory for Threatening Social Actions," *Journal of Experimental Child Psychology* 112, no. 1 (2012): 102–110.

210 **a romantic relationship must be marked:** Ellie Lisitsa, "The Positive Perspective: Dr. Gottman's Magic Ratio!" Gottman Institute, December 5, 2012, https://www.gottman.com/blog/the-positive-perspective-dr-gottmans-magic-ratio.

210 **the worse the action, the more likely:** Susan T. Fiske, "Attention and Weight in Person Perception: The Impact of Negative and Extreme Behavior," *Journal of Personality and Social Psychology* 38 (1980): 889–906.

210 **And so, by some estimates:** Ray Williams, "Why We Love Bad News More Than Good News," *Psychology Today*, November 1, 2014, https://www.psychologytoday.com/blog/wired-success/201411/why-we-love-bad-news-more-good-news; Marc Trussler and Stuart Soroka, "Consumer Demand for Cynical and Negative News Frames," *International Journal of Press/Politics* 19, no. 3 (2014): 360–379.

211 **the mistaken but common belief that the world is dangerous:** Moran Bodas, Maya Siman-Tov, Kobi Peleg, and Zahava Solomon, "Anxiety-Inducing Media: The Effect of Constant News Broadcasting on the Well-being of Israeli Television

Viewers," *Psychiatry* 78, no. 3 (2015): 265–276; Sean Patrick Roche, Justin T. Pickett, and Marc Gertz, "The Scary World of Online News? Internet News Exposure and Public Attitudes Toward Crime and Justice," *Journal of Quantitative Criminology* 32, no. 2 (2016): 215–236; Sara Tiegreen and Elana Newman, "Violence: Comparing Reporting and Reality," Dart Center for Journalism & Trauma, February 18, 2009, http://dartcenter.org/content/violence-comparing-reporting-and-reality.

211 The word "epidemic" frequently crops up: Peter Moore, "Does America Have a Rape Culture?" YouGovUS, December 11, 2014, https://today.yougov.com/news/2014/12/11/rape-culture.

211 rates of sexual assault are *decreasing,* not increasing: "Yes Means Yes, Says Mr. Brown," *The Economist,* October 3, 2014; Sofi Sinozich and Lynn Langton, "Rape and Sexual Assault Among College-Age Females, 1995–2013," NCJ 248471 (Washington, DC: Bureau of Justice Statistics, December 11, 2014), http://www.bjs.gov/index.cfm?ty=pbdetail&iid=5176.

212 About half of the respondents reported: Common Cause Foundation, *Perceptions Matter: The Common Cause UK Values Survey* (London: Common Cause Foundation, 2016). The findings of the survey are also broadly consistent with findings of the recent laboratory study by Ben M. Tappin and Ryan T. McKay, "The Illusion of Moral Superiority," *Social Psychological and Personality Science* (2016): 1–9.

213 Think, for example, of Anne Frank: Anne Frank, *The Diary of Anne Frank* (New York: Doubleday, 2001), 333; "In Our Opinion: Nelson Mandela Left Legacy of Freedom and Faith" (editorial), *Deseret News,* December 5, 2013; Mazhar Kibriya, *Gandhi and Indian Freedom Struggle* (New Delhi: APH Publishing, 1999), 20; Martin Luther King Jr., Nobel acceptance speech, December 10, 1964, Nobelprize.org, https://www.nobelprize.org/nobel_prizes/peace/laureates/1964/king-acceptance_en.html. A little later in King's speech, he predicted that goodness would be the rule of the land when "the lion and the lamb shall lie down together." And as you have seen, they have.

213 the beliefs of Richard Ramirez, the notorious serial murderer: Vojtech Mastny, *The Cold War and Soviet Insecurity: The Stalin Years* (New York: Oxford University Press, 1996), 128; Mark Thomas, *The Deadliest War: The Story of World War II* (Berlin, NJ: Townsend Press, 2011); *International Encyclopedia of Public Health,* vol. 1, 2nd ed., edited by Stella R. Quah and William Cockerham (Waltham, MA: Elsevier), 467.

215 Starting from an assumption of others' trustworthiness: Robert M. Axelrod, *The Evolution of Cooperation* (New York: Basic Books, 2006). The relationship between trust and cooperation is undeniably complex. Just as trust can lead to cooperation, so can cooperation lead to trust; see Alexander Peysakhovich and David G. Rand, "Habits of Virtue: Creating Norms of Cooperation and Defection in the Laboratory," *Management Science* 62, no. 3 (2013): 631–647.

216 In one computer simulation study we conducted: Kristin M. Brethel-Haurwitz, Sarah A. Stoycos, Elise M. Cardinale, Bryce Huebner, and Abigail A. Marsh, "Is Costly Punishment Altruistic? Exploring Rejection of Unfair Offers in the Ultimatum Game in Real-World Altruists," *Scientific Reports* 6 (2016): 18974.

217 the *social discounting* task: Bryan Jones and Howard Rachlin, "Social Discounting," *Psychological Science* 17, no. 4 (2006): 283–286.

218 This pattern holds up across multiple studies: Howard Rachlin and Bryan A. Jones, "Social Discounting and Delay Discounting," *Journal of Behavioral Decision Making* 21 (2008): 29–43; Qingguo Ma, Guanxiong Pei, and Jia Jin, "What

Makes You Generous? The Influence of Rural and Urban Rearing on Social Discounting in China," *PLoS One* 10, no. 7 (2015): e0133078; Tina Strombach, Jia Jin, Bernd Weber, Peter H. Kenning, Qiang Shen, Qingguo Ma, and Tobias Kalenscher, "Charity Begins at Home: Cultural Differences in Social Discounting and Generosity," *Journal of Behavioral Decision Making* 27, no. 3 (2014): 235–245.

220 **Their generosity had dropped by less than half:** Kruti M. Vekaria, Kristin M. Brethel-Haurwitz, Elise M. Cardinale, Sarah A. Stoycos, and Abigail A. Marsh, "Social Discounting and Distance Perceptions in Costly Altruism," *Nature Human Behaviour* (2017).

221 **as the philosopher Peter Singer and others have put it:** Peter Singer, *The Expanding Circle: Ethics and Sociobiology* (Oxford: Clarendon Press/Oxford University Press, 1981).

222 **Reciprocal altruism is the closest:** Gabriele Bellucci, Sergey V. Chernyak, Kimberly Goodyear, Simon B. Eickhoff, and Frank Krueger, "Neural Signatures of Trust in Reciprocity: A Coordinate-Based Meta-Analysis," *Human Brain Mapping* (2016), DOI: 10.1002/hbm.23451; James K. Rilling and Alan G. Sanfey, "The Neuroscience of Social Decision-Making," *Annual Review of Psychology* 62 (2011): 23–48.

222 **altruistic kidney donors overwhelmingly report:** Sadler et al., "The Living, Genetically Unrelated, Kidney Donor"; Stothers, Gourlay, and Liu, "Attitudes and Predictive Factors for Live Kidney Donation."

223 **My colleague David Rand, a behavioral scientist:** David G. Rand, "Cooperation, Fast and Slow: Meta-Analytic Evidence for a Theory of Social Heuristics and Self-Interested Deliberation," *Psychological Science* 27, no. 9 (2016): 1192–1206.

224 **urges to care and cooperate are deeply rooted:** David G. Rand and Ziv G. Epstein, "Risking Your Life Without a Second Thought: Intuitive Decision-Making and Extreme Altruism," *PLoS One* 9, no. 10 (2014): e109687.

225 **its advocates' explicit aim:** From the website effectivealtruism.org: "Effective altruism is changing the way we do good. Effective altruism is about answering one simple question: how can we use our resources to help others the most? Rather than just doing what feels right, we use evidence and careful analysis to find the very best causes to work on."

225 **quickly spiral into a vortex of indecision:** For thoughtful critiques of effective altruism, see Dylan Matthews, "I Spent a Week at Google Talking with Nerds About Charity. I Came Away . . . Worried," *Vox,* August 10, 2015, http://www.vox.com/2015/8/10/9124145/effective-altruism-global-ai; Eric Posner, "Should Charity Be Logical?" *Slate,* March 26, 2015, http://www.slate.com/articles/news_and_politics/view_from_chicago/2015/03/effective_altruism_critique_few_charities_stand_up_to_rational_evaluation.html; Jamil Zaki, "The Feel-Good School of Philanthropy," *New York Times,* December 5, 2015.

226 **Such injuries leave IQ and reasoning ability intact:** Antonio R. Damasio, *Descartes' Error* (New York: Random House, 2006); Edmund T. Rolls and Fabien Grabenhorst, "The Orbitofrontal Cortex and Beyond: From Affect to Decision-Making," *Progress in Neurobiology* 86, no. 3 (2008): 216–244.

226 **"There is a wholly fallacious theory":** Bertrand Russell, "What Desires Are Politically Important?" (Nobel lecture), December 11, 1950, nobelprize.org, http://www.nobelprize.org/nobel_prizes/literature/laureates/1950/russell-lecture.html.

226 **Consider the case of Robert Mather:** Derek Thompson, "The Greatest Good," *The Atlantic,* June 15, 2015.

229 **the amount of suffering and misery:** For graphs showing these changes and many others, see Max Roser, "Life Expectancy," Our World in Data, https://ourworldindata.org/life-expectancy.

229 **The economic historian Joel Mokyr:** Ana Swanson, "Why the Industrial Revolution Didn't Happen in China," Washington Post, October 28, 2016.

229 **The World Bank estimates**: The World Bank, "World Monitoring Report 2015/2016," http://www.worldbank.org/en/publication/global-monitoring-report.

230 **We ran a number of analyses to probe:** Kristin M. Brethel-Haurwitz and Abigail A. Marsh, "Geographical Differences in Subjective Well-being Predict Extraordinary Altruism," *Psychological Science* 25 (2014): 762–771.

232 **The psychologists Elizabeth Dunn and Mike Norton:** Lara B. Aknin, Christopher P. Barrington-Leigh, Elizabeth W. Dunn, John F. Helliwell, Justine Burns, Robert Biswas-Diener, Imelda Kemeza, Paul Nyende, Claire E. Ashton-James, and Michael I. Norton, "Prosocial Spending and Well-being: Cross-Cultural Evidence for a Psychological Universal," *Journal of Personality and Social Psychology* 104, no. 4 (2013): 635–652; Lara B. Aknin, Elizabeth W. Dunn, and Michael I. Norton, "Happiness Runs in a Circular Motion: Evidence for a Positive Feedback Loop Between Prosocial Spending and Happiness," *Journal of Happiness Studies* 13 (2012): 347–355.

232 **One 2005 Gallup poll found a linear relationship:** Carroll, "Americans More Likely to Donate Money"; see also Jesus Ramirez-Valles, "Volunteering in Public Health: An Analysis of Volunteers' Characteristics and Activities," *International Journal of Volunteer Administration* 24, no. 2 (2006): 15–24.

232 **Another large study found similar results:** Paul B. Reed and L. Kevin Selbee, "The Civic Core in Canada: Disproportionality in Charitable Giving, Volunteering, and Civic Participation," *Nonprofit and Voluntary Sector Quarterly* 30, no. 4 (2001): 761–780.

232 **A field experiment in Ireland:** Antonio S. Silva and Ruth Mace, "Cooperation and Conflict: Field Experiments in Northern Ireland," *Proceedings of the Royal Society B: Biological Sciences* 281, no. 1792 (2014): 20141435; Jo Holland, Antonio S. Silva, and Ruth Mace, "Lost Letter Measure of Variation in Altruistic Behaviour in 20 Neighbourhoods," *PLoS One* 7, no. 8 (2012): e43294; David Sloan Wilson, Daniel Tumminelli O'Brien, and Artura Sesma, "Human Prosociality from an Evolutionary Perspective: Variation and Correlations at a City-Wide Scale," *Evolution and Human Behavior* 30, no. 3 (2009): 190–200.

233 **The positive relationship between well-being and altruism:** Sebastian Prediger, Björn Vollan, and Benedikt Herrmann, "Resource Scarcity, Spite and Cooperation," Working Papers in Economics and Statistics 2013-10, University of Innsbruck, 2013; Yu-Kang Lee and Chun-Tuan Chang, "Who Gives What to Charity? Characteristics Affecting Donation Behavior," *Social Behavior and Personality: An International Journal* 35, no. 9 (2007): 1173–1180.

233 **Many decades of research on misanthropy:** Tom W. Smith, "Factors Relating to Misanthropy in Contemporary American Society," *Social Science Research* 26, no. 2 (1997): 170–196.

233 **To be fair, some recent studies:** Paul K. Piff, Daniel M. Stancato, Stéphane Côté, Rodolfo Mendoza-Denton, and Dacher Keltner, "Higher Social Class Predicts Increased Unethical Behavior," *Proceedings of the National Academy of Sciences United States of America* 109, no. 11 (2012): 4086–4091; Paul K. Piff, Michael W. Kraus, Stéphane Côté, Bonnie Hayden Cheng, and Dacher Keltner, "Having

Less, Giving More: The Influence of Social Class on Prosocial Behavior," *Journal of Personality and Social Psychology* 99, no. 5 (2010): 771–84.

234 The study was conducted by the German psychologist: Martin Korndörfer, Boris Egloff, and Stefan C. Schmukle, "A Large Scale Test of the Effect of Social Class on Prosocial Behavior," *PLoS One* 10, no. 7 (2015): e0133193.

235 despite the fact that wealthier individuals: Derek Thompson, "The Free-Time Paradox in America," *The Atlantic,* September 13, 2016; "Why Is Everyone So Busy?" *The Economist,* December 20, 2015.

240 Prosperity within a culture tends to be associated: Patricia M. Greenfield, "The Changing Psychology of Culture from 1800 Through 2000," *Psychological Science* 24, no. 9 (2013): 1722–1731; Pamela L. Cox, Barry A. Friedman, and Thomas Tribunella, "Relationships Among Cultural Dimensions, National Gross Domestic Product, and Environmental Sustainability," *Journal of Applied Business and Economics* 12, no. 6 (2011): 46; Ronald Inglehart and Wayne E. Baker, "Modernization, Cultural Change, and the Persistence of Traditional Values," *American Sociological Review* (2000): 19–51; Pinker, *Better Angels of Our Nature.*

240 Collectivism entails focusing on and valuing interdependence: Nicholas Sorensen and Daphna Oyserman, "Collectivism, Effects on Relationships," May 2, 2011, in *Encyclopedia of Human Relationships* (Sage Publications, 2009), http://dornsife.usc.edu/assets/sites/783/docs/sorensen_oyserman_2009.pdf.

240 Analyses of cultural values across nations: Cox, Friedman, and Tribunella, "Relationships Among Cultural Dimensions," 46; Linghui Tang and Peter E. Koveos, "A Framework to Update Hofstede's Cultural Value Indices: Economic Dynamics and Institutional Stability," *Journal of International Business Studies* 39, no. 6 (2008): 1045–1063; Henri C. Santos, Michael E. W. Varnum, and Igor Grossmann, "Global Increases in Individualism," *Psychological Science* (2017): in press.

240 one recent study found that over the last 200 years: Greenfield, "The Changing Psychology of Culture from 1800 Through 2000."

240 the recent period of incredible economic growth in China: Liza G. Steele and Scott M. Lynch, "The Pursuit of Happiness in China: Individualism, Collectivism, and Subjective Well-being During China's Economic and Social Transformation," *Social Indicators Research* 114, no. 2 (2013): 441–451.

240 This effect may be bidirectional: Igor Grossmann and Michael E. W. Varnum, "Social Structure, Infectious Diseases, Disasters, Secularism, and Cultural Change in America," *Psychological Science* 26, no. 3 (2015): 311–324; Yuriy Gorodnichenko and Gérard Roland, "Individualism, Innovation, and Long-Run Growth," *Proceedings of the National Academy of Sciences United States of America* 108, suppl. 4 (2011): 21316–21319.

240 traditional Confucian teachings emphasize: Wilbur A. Lam and Laurence B. McCullough, "Influence of Religious and Spiritual Values on the Willingness of Chinese-Americans to Donate Organs for Transplantation," *Clinical Transplantation* 14, no. 5 (2000): 449–456; Andrew Ma, "Comparison of the Origins of Altruism as Leadership Value Between Chinese and Christian Cultures," *Leadership Advance Online* XVI (2009).

241 "Fijians do games that involve giving": Quoted in Bob Holmes, "Generous by Nature," *New Scientist* 231, no. 3086 (2016): 26–28.

241 An emphasis on group bonds requires: Anu Realo, Jüri Allik, and Brenna Greenfield, "Radius of Trust: Social Capital in Relation to Familism and Institutional Collectivism," *Journal of Cross-Cultural Psychology* 39, no. 4 (2008):

447–462; Yuriy Gorodnichenko and Gérard Roland, "Understanding the Individualism-Collectivism Cleavage and Its Effects: Lessons from Cultural Psychology," in *Institutions and Comparative Economic Development,* edited by Masahiko Aoki, Timur Kuran, and Gérard Roland, 213–236 (Berlin: Springer, 2012); André Van Hoorn, "Individualist-Collectivist Culture and Trust Radius: A Multilevel Approach," *Journal of Cross-Cultural Psychology* 46, no. 2 (2015): 269–276. For interesting real-world implications, see "The Unkindness of Strangers," *The Economist,* July 27, 2013.

241 **Collectivism is associated with low levels:** Mie Kito, Masaki Yuki, and Robert Thomson, "Relational Mobility and Close Relationships: A Socioecological Approach to Explain Cross-Cultural Differences," *Personal Relationships* 24, no. 1 (2017): 114–130; Masaki Yuki and Joanna Schug, "Relational Mobility: A Socioecological Approach to Personal Relationships," in *Relationship Science: Integrating Evolutionary, Neuroscience, and Sociocultural Approaches,* edited by Omri Gillath, Glenn Adams, and Adrianne Kunkel, 137–151 (Washington DC: American Psychological Association, 2012).

242 **when the wider culture paints members of an outgroup:** Brad Pinter and Anthony G. Greenwald, "A Comparison of Minimal Group Induction Procedures," *Group Processes Intergroup Relations* 14 (2011): 81–98; Mina Cikara, Emile G. Bruneau, and Rebecca R. Saxe, "Us and Them: Intergroup Failures of Empathy," *Current Directions in Psychological Science* 20 (2011): 149–153; Jonathan Levy, Abraham Goldstein, Moran Influs, Shafiq Masalha, Orna Zagoory-Sharon, and Ruth Feldman, "Adolescents Growing Up Amidst Intractable Conflict Attenuate Brain Response to Pain of Outgroup," *Proceedings of the National Academy of Sciences United States of America* 113, no. 48 (2016): 13696–13701.

242 **These psychological phenomena may help to explain:** Gil Luria, Ram Cnaan, and Amnon Boehm, "National Culture and Prosocial Behaviors Results from 66 Countries," *Nonprofit and Voluntary Sector Quarterly* (2014): 0899764014554456.

242 **The top of this index is reliably dominated:** "Individualism," Clearly Cultural, http://www.clearlycultural.com/geert-hofstede-cultural-dimensions/individualism/.

242 **These countries are also vastly more likely:** Alois Gratwohl, Helen Baldomero, Michael Gratwohl, Mahmoud Aljurf, Luis Fernando Bouzas, Mary Horowitz, Yoshihisa Kodera, Jeff Lipton, Minako Iida, Marcelo C. Pasquini, Jakob Passweg, Jeff Szer, Alejandro Madrigal, Karl Frauendorfer, Dietger Niederwieser, and WBMT (Worldwide Network of Blood and Marrow Transplantation), "Quantitative and Qualitative Differences in Use and Trends of Hematopoietic Stem Cell Transplantation: A Global Observational Study," *Haematologica* 98, no. 8 (2013): 1282–1290; WHO, "Blood Safety and Availability"; GODT (Global Observatory on Donation and Transplantation), "WHO-ONT," http://www.transplant-observatory.org/.

243 **This may contribute to the persistent problem:** A. L. N. Udegbe, Kemi Ololade Odukoya, and Babatunde E. Ogunnowo, "Knowledge, Attitude, and Practice of Voluntary Blood Donation Among Residents in a Rural Local Government Area in Lagos State: A Mixed Methods Survey," *Nigerian Journal of Health Sciences* 15, no. 2 (2015): 80; S. O. Onuh, M. C. Umeora, and Odidika Ugochukwu Joannes Umeora, "Socio-Cultural Barriers to Voluntary Blood Donation for Obstetric Use in a Rural Nigerian Village," *African Journal of Reproductive Health* (2005): 72–76; Osaro Erhabor and Teddy Charles Adias, "The Challenges of Meeting the

Blood Transfusion Requirements in Sub-Saharan Africa: The Need for the Development of Alternatives to Allogenic Blood," *Journal of Blood Medicine* 2 (2011): 7–21; Anju Dubey, Atul Sonker, Rahul Chaurasia, and Rajendra Chaudhary, "Knowledge, Attitude, and Beliefs of People in North India Regarding Blood Donation," *Blood Transfusion* 12, suppl. 1 (2014): s21–s27; Tanja Z. Zanin, Denise P. Hersey, David C. Cone, and Pooja Agrawal, "Tapping into a Vital Resource: Understanding the Motivators and Barriers to Blood Donation in Sub-Saharan Africa," *African Journal of Emergency Medicine* 6, no. 2 (2016): 70–79. For a particularly thorough exploration of relevant factors, see the following ethnographic study conducted in Pakistan: Zubia Mumtaz, Sarah Bowen, and Rubina Mumtaz, "Meanings of Blood, Bleeding, and Blood Donations in Pakistan: Implications for National vs. Global Safe Blood Supply Policies," *Health Policy and Planning* 27, no. 2 (2012): 147–155..

243 **Even within an individualist nation like the United States:** Markus Kemmelmeier and Joyce A. Hartje, "Individualism and Prosocial Action: Cultural Variations in Community Volunteering," *Advances in Psychology Research* 51 (2007): 149; Lam and McCullough, "Influence of Religious and Spiritual Values."

244 **Fiction, in particular, represents:** Keith Oatley, "Fiction: Simulation of Social Worlds," *Trends in Cognitive Sciences* 20, no. 8 (2016): 618–628.

244 **some subjects read a brief note:** C. Daniel Batson and Nadia Y. Ahmad, "Empathy-Induced Altruism in a Prisoner's Dilemma II: What If the Target of Empathy Has Defected?" *European Journal of Social Psychology* 31, no. 1 (2001): 25–36.

245 **People who read more fiction:** Oatley, "Fiction: Simulation of Social Worlds"; P. Matthjis Bal and Martijn Veltkamp, "How Does Fiction Reading Influence Empathy? An Experimental Investigation on the Role of Emotional Transportation," *PLoS One* 8, no. 1 (2013): e55341.

Chapter 8: Putting Altruism into Action

247 **Collectivist cultures generally value conformity:** Markus Kemmelmeier and Joyce A. Hartje, "Individualism and Prosocial Action: Cultural Variations in Community Volunteering," *Advances in Psychology Research* 51 (2007): 149.

247 **Yamagishi has proposed that this may explain:** Toshio Yamagishi and Midori Yamagishi, "Trust and Commitment in the United States and Japan," *Motivation and Emotion* 18, no. 2 (1994): 129–166.

248 **An influential series of studies:** Netta Weinstein and Richard M. Ryan, "When Helping Helps: Autonomous Motivation for Prosocial Behavior and Its Influence on Well-being for the Helper and Recipient," *Journal of Personality and Social Psychology* 98, no. 2 (2010): 222–244.

249 **the evident satisfaction felt by Lenny Skutnik:** "Hero of Plane Crash Had Little Experience in the Hero Business," *Los Angeles Times/Washington Post* News Service, January 16, 1982.

249 **the confusion of foreseen outcomes with intended outcomes:** Thomas A. Cavanaugh, "The Intended/Foreseen Distinction's Ethical Relevance," *Philosophical Papers* 25, no. 3 (1996): 179–188.

250 **As the Buddhist monk and neuroscience researcher:** Matthieu Ricard, *Altruism: The Power of Compassion to Change Yourself and the World* (London: Atlantic Books, 2015), 141.

250 **Expending resources on helping others:** Elizabeth W. Dunn, Lara B. Aknin, and Michael I. Norton, "Spending Money on Others Promotes Happiness," *Science* 319, no. 5870 (2008): 1687–1688.

251 **Once a mother rat has had the experience:** Cort Andrew Pedersen, "Biological Aspects of Social Bonding and the Roots of Human Violence," *Annals of the New York Academy of Sciences* 1036 (2004): 106–127.

251 **the amygdala lesion patient S.M. is not a psychopath:** Scott O. Lilienfeld, Katheryn C. Sauvigné, Justin Reber, Ashley L. Watts, Stephan B. Hamann, Sarah Francis Smith, Christopher J. Patrick, Shauna M. Bowes, and Daniel Tranel, "Potential Effects of Severe Bilateral Amygdala Damage on Psychopathic Personality Features: A Case Report," *Personality Disorders: Theory, Research, and Treatment* (December 2016), DOI: 10.1037/per0000230.

251 **One recent tantalizing study:** Robin S. Rosenberg, Shawnee L. Baughman, and Jeremy N. Bailenson, "Virtual Superheroes: Using Superpowers in Virtual Reality to Encourage Prosocial Behavior," *PLoS One* 8, no. 1 (2013): e55003.

251 **a twenty-year-old program with demonstrated success:** Kimberly A. Schonert-Reichl, Veronica Smith, Anat Zaidman-Zait, and Clyde Hertzman, "Promoting Children's Prosocial Behaviors in School: Impact of the 'Roots of Empathy' Program on the Social and Emotional Competence of School-Aged Children," *School Mental Health* 4, no. 1 (2012): 1–21.

252 **Even relatively brief training in compassion:** Daniel Lim, Paul Condon, and David DeSteno, "Mindfulness and Compassion: An Examination of Mechanism and Scalability," *PLoS One* 10, no. 2 (2015): e0118221; Paul Condon, Gaëlle Desbordes, Willa Miller, and David DeSteno, "Meditation Increases Compassionate Responses to Suffering," *Psychological Science* 24 (2013): 2125–2127; Julieta Galante, Marie-Jet Bekkers, Clive Mitchell, and John Gallacher, "Loving-Kindness Meditation Effects on Well-being and Altruism: A Mixed-Methods Online RCT," *Applied Psychology: Health and Well-being* (2016); Yoona Kang, Jeremy R. Gray, and John F. Dovidio, "The Nondiscriminating Heart: Lovingkindness Meditation Training Decreases Implicit Intergroup Bias," *Journal of Experimental Psychology General* 143, no. 3 (2014): 1306–1313.

254 **Humility, happily, is one of those rare and wonderful qualities:** Joshua D. Foster, W. Keith Campbell, and Jean M. Twenge, "Individual Differences in Narcissism: Inflated Self-Views Across the Lifespan and Around the World," *Journal of Research in Personality* 37, no. 6 (2003): 469–486; Petar Milojev and Chris G. Sibley, "The Stability of Adult Personality Varies Across Age: Evidence From a Two-Year Longitudinal Sample of Adult New Zealanders," *Journal of Research in Personality* 51 (2014): 29–37.

254 **"Just driving in our car":** K. K. Ottesen, "Cory Booker on the Perils of Heroism," *Washington Post*, February 25, 2016.

INDEX

Abigail Marsh is an associate professor of psychology at Georgetown University, where she directs the Laboratory on Social and Affective Neuroscience. She received her PhD in social psychology from Harvard University. Her work has been covered on NPR and in the *Wall Street Journal, The Economist, Slate, Huffington Post,* and *New York Magazine;* she also writes a blog for *Psychology Today* and presented her research on altruism in a 2016 TED Talk. Her lab's work on extraordinary altruism was awarded the Cozzarelli Prize by the *Proceedings of the National Academy of Sciences*. She lives in Washington, DC.

http://www.abigailmarsh.com

Photography by Phoebe Taubman